GLEB KOSORUKOV

HEROES OF LABOUR

STEIDL

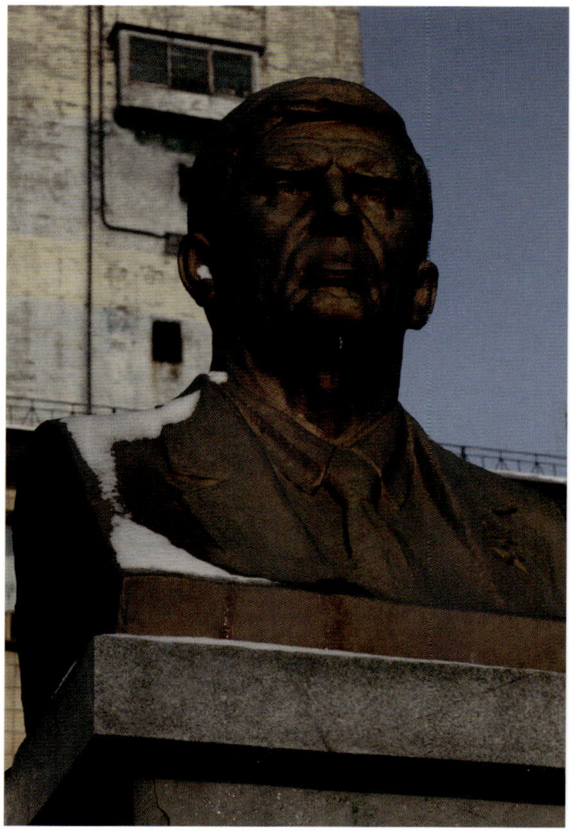

The *Heroes of Labour* book and installation is an attempt to question radical changes in the very notion of labour in contemporary, post-industrial society.

The global balance of power, which once helped to support a certain level of social justice in capitalist society, as each side battled for the moral high ground, shifted with the demise of the Soviet socialist system. Brute capitalism was then unleashed on an unsuspecting global stage. Entire industries are now being moved to parts of the world where the workforce is cheap and plentiful, countries where workers can be hired and fired at will. The proletariat, as defined by Marxist political economics, is gradually disappearing and being replaced by a new class – the precariat, a precarious, temporary and ever changeable workforce. Labour itself is losing value in this era of capitalist globalisation. While working on *Heroes of Labour* I also appealed to the past – a time when the value of labour was at its zenith.

On 31 August 1935 Alexey Stakhanov, a jackhammer operator at the Central Irmino coal mine, mined a record 102 tonnes of coal in 5 hours and 45 minutes (fourteen times his daily quota). The launch of an unprecedented state-run campaign for popularising extraordinary labour achievements made Stakhanov a Soviet hero par excellence. Soon after, his portrait appeared on the cover of *Time* magazine. For the first time, a worker had been elevated to worldwide fame for his performance at work. Stakhanov, by virtue of history, was destined to turn labour into a matter of heroism.

Seventy-four years later I started a research project on miners' identities at the largest coalface in Europe, aptly named the Stakhanov mine. This mine, with seams ranging from 830 to 1,240 metres deep, was opened in 1974 near the city of Krasnoarmiisk in the Donetsk region in the south-east Ukraine. Today, its glory days long since over, the Stakhanov mine produces six times less coal than it did at the end of the Soviet era. The onward march of global capitalism and tectonic changes in the labour markets mean that Ukranian mines are closing apace. Up to 100,000 miners will lose their jobs within the next five years. With all this in mind, my aim and wish was to examine the last vestige of Stakhanov's legacy through the image of workers today.

Since the moment I started working on the project, much has changed in the south-east Ukraine. Miners are not only losing their jobs, but their lives, as participants and victims in the war for independence in the Donetsk and Lugansk republics.

Gleb Kosorukov

Donetsk, 2015

In the past, multimedia artist Gleb Kosorukov has worked for major international news and fashion publications. He spends much of his time in Paris and feels at home in the international multicultural environment. However, his heart belongs to the people of a very different world. One of his recent projects is devoted to the very first and perhaps the last true proletarians – miners, whom he documented in the south-east Ukraine.

The term "working class" has distinctly different connotations in the West compared to those it had in the Soviet Union. In the West the words have a historical and romantic feel, laced with the mythology of class struggles of the past. This very mythology has been undermined by socio-economic progress along with the rapid erosion of the industrial proletariat, accelerated by technological and social changes in the global capitalist economy. The discourse of the left operating with neo-Marxist ideas remains largely disconnected from the classical Marxist definition of the proletariat. And yet Baudrillard's maxims such as "the proletariat works in order to strike" did have some relation to reality.

The proletariat of the Soviet Union had a very special status. The actual life of the proletariat was, as with every other layer of this society, controlled by government officials. However the ideological hegemony of the working class, as the driving force of society, was entirely preserved. Industrial labour was certainly honourable, often regardless of its efficiency and functional necessity. Stakhanov, the record-holding miner, famously produced 102 tonnes of ore in 5 hours and 45 minutes in 1935, and continued as late as 1970 to be a hero and role model – although to many of his contemporaries it was clear even at the time that this kind of triumphant (and possibly staged) stunt had much to do with propaganda.

In art, the theme of the proletariat was prioritised and privileged and, from 1930 onwards, it dominated the artistic canon. In time, specific nuances were incorporated (so that in the Khrushchev "thaw" the so-called stricter style was introduced, articulating the intricacies of the working day, in contrast to the official pomp of labour, etc.), but in general the theme itself was unshakable.

For thoughtful independent artists, the working motif seemed officious and servile. The collapse of the Soviet Union put an end to the prioritisation of labour as the chief artistic theme. Belonging to the working class, devoid of the mythological camouflage, lost its attraction and exposed its social vulnerability, lack of mobility, hopelessness, uselessness. In the 2006 movie *The Last Strike*, the heroes, miners of different generations, plan to blow themselves up at the bottom of the shaft to draw attention to the plight of their fellow men.

Art and the proletariat parted ways. Workers appeared as a class only in the generalised discourse of the left and not as images of flesh and blood; they became simply a concept. In this decidedly bleak situation Kosorukov's work demonstrates loyalty to the vanishing class, to real living people, not symbolic constructs.

In 2001, together with Alexei Guintovt and Andrei Molodkin, Kosorukov became involved in a project called *Novo Novosibirsk* at the State Russian Museum. It was a complex artistic undertaking, representing dreams of the ruins of an empire in a grand style. For this Kosorukov created photographic images that signalled the Soviet photographic avant-garde. These references played with our contact with reality – hence the difficult angles, diagonal constructions, accented monumentality and so forth. It was like he was visualising mythology of the utopian neo-proletarian state.

In another project, *Procession*, his perception was immersed in the reality that is life, according to the vocabulary of classical realist art. Kosorukov clearly reflects on polyphonic narratives in Russian democratic art of the second half of the nineteenth century, the processions of Vasily Perov, Ilya Repin, Illarion Pryanishnikov. The social criticism of those artists always coexisted with a poetic, perhaps prophetic, sense of history. Kosorukov does not stylise these textbook images – he simply feels this legacy behind him.

He is ruthless in this series, but also lyrical and palpably anxious. As part of his process he grasped the archetypes of modern Russia, their social standing (including the taboo subject of Soviet art – the lower classes), their hopes and fears. This series is Kosorukov's response to the recent top-down approach in search for a new national ideology, a collective identity (not only in Russia but also in neighbouring states). In these images a procession of archetypes takes us through the ruins of what was once Europe's largest high-technology glass factory where the ruby stars were made for the Kremlin's towers as well as the transparent coffin that contains Lenin's preserved body at his mausoleum in Red Square. The second hint, revealing the mobilisation of the identity search, was a video performance where the author with his competitor carried out an endless stunt of consecutive assembling-disassembling the iconic Kalashnikov machine gun, while a chess clock measured the elapsed time.

In Kosorukov's depiction of the working class he is almost alone. Before him Arsen Savadov produced the partly anarchist, partly dadaist series *Donbass-Chocolate*. Miners were the pride of the Soviet Union, as was the classical ballet, so to dress the miners up in tutus was to dance on the bones of the USSR. Boris Mikhailov, great photographer of the older generation, was always attentive to the physiology of modern life, to social nuance, the homeless, and so on. Later he came to images of popular revolt, interpreted in the *Maidan* series with the physiology of Ilya Repin's *Barge Haulers on the Volga*.

Kosorukov treads a different path in his project *The Heroes of Labour*. Employing a researcher's methodology, he bypasses a narrative or manipulative approach, executing a statistically significant number of portraits of randomly chosen miners at a particular place (the Stakhanov mine) in a short period of time. Measuring human value with his camera, he shows miners as they really are. Forced to work in an increasingly hazardous environment for pennies, deprived of their former status, nearly defunct and almost redundant, like ever before they are again a combustible material for social conflicts.

The original title of his portrait series was *The Stakhanov's Hundred* (a play on words since "Stakhanov" is the name of the mine where the portraits were taken), and perhaps an ironic reference to the mythical figure and heroic symbol of Stalin's Five Year Plan. However, there is little irony in Kosorukov's visual language – the author does not knock a man when he is down, and remains true to his subjects.

This portrait series became a core of the installation *Heroes of Labour*, which originally occupied an entire floor at the Waterschei mine in Genk, Belgium (the building which hosted Manifesta 9, the European Biennial of Contemporary Art, the year before). In this vast space more than a hundred large-format photos were lit intermittently by industrial projectors controlled by motion detection sensors. When the viewer stopped moving, he or she was left in complete darkness. This was not a recreation of the mining environment (in Europe there are mining museums which create the illusion of actually being under the ground); it was a way of putting the viewer one-on-one with each of the miners. Enlargement here is not monumental (a technique often used to highlight labour themes in the Soviet Union), but a way to establish the close connection with each person photographed. The artist insists that his characters are physical, corporeal, real.

I do not know whether Kosorukov is under the spell of some leftist discourse, but as an artist, rooted in a certain material life (the working and living environment of his parents, peers and neighbours), he focuses solely on authenticity. No manipulation. We feel he knows his characters inside out – knows their hopes, fears and needs. Perhaps this is the right word – needs. Lately in the field of contemporary art (even on the level of research) we always meet the idea of the proletariat and what it stands for or embodies as a product of neo-Marxist theory, not what it is, but an intellectual concept. Here, the artist positions himself firmly in reality, conducting statistically viable, randomised visual research.

In one of the three video installations we see a video-portrait of a woman singing *The International* – the battle hymn of the proletariat on the rise of its historical and world-transforming mission. Why is it here? The irony of the current state of the proletariat, its mutations, the state of defeat? The irony of the new left-wing intellectuals, constantly coming up with new theories? Again, there is nothing ironic in his video. The hymn once created for the chorus is sung by a single individual, an unemployed ex-proletarian woman, busking on a suburban train. The singer has the nervous face of the intellectual, the inspired face of a human. Is it a farewell song to the working class leaving the stage of history, along with the classical Marxist ideas of mass production? Is it referencing this class in its current condition, fighting its last battle? I do not know. But I do know that it is with respect that the author addresses this theme.

It stems from the moment when social science and art looked in the face of the worker with respect, and started the period in history which was later called "the victorious march of the proletariat". From the middle of last century the image of the working class in art became more stilted, conditional, canonical. Kosorukov takes us back to a time when the proletariat was seen humanely, compassionately, as a community of oppressed people. He is attentive to the simple needs of its heroes, lending those old notions of oppression and solidarity a real sympathy – a rare quality that seems for evermore to be buried under the multiple layers of pompous art and art mediated by social theories.

Alexander Borovsky

Head of the Department of
Contemporary Art at the State Russian Museum

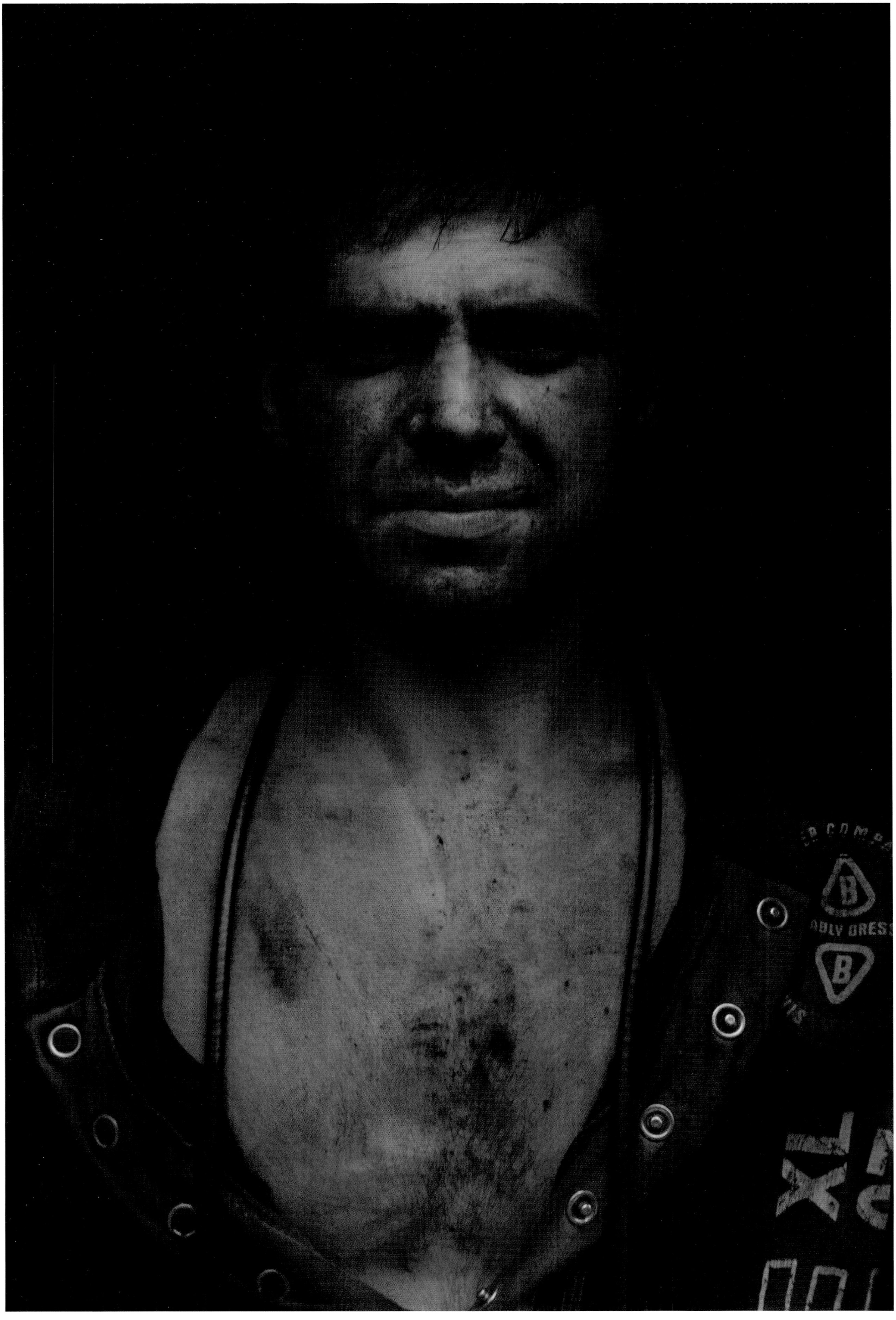

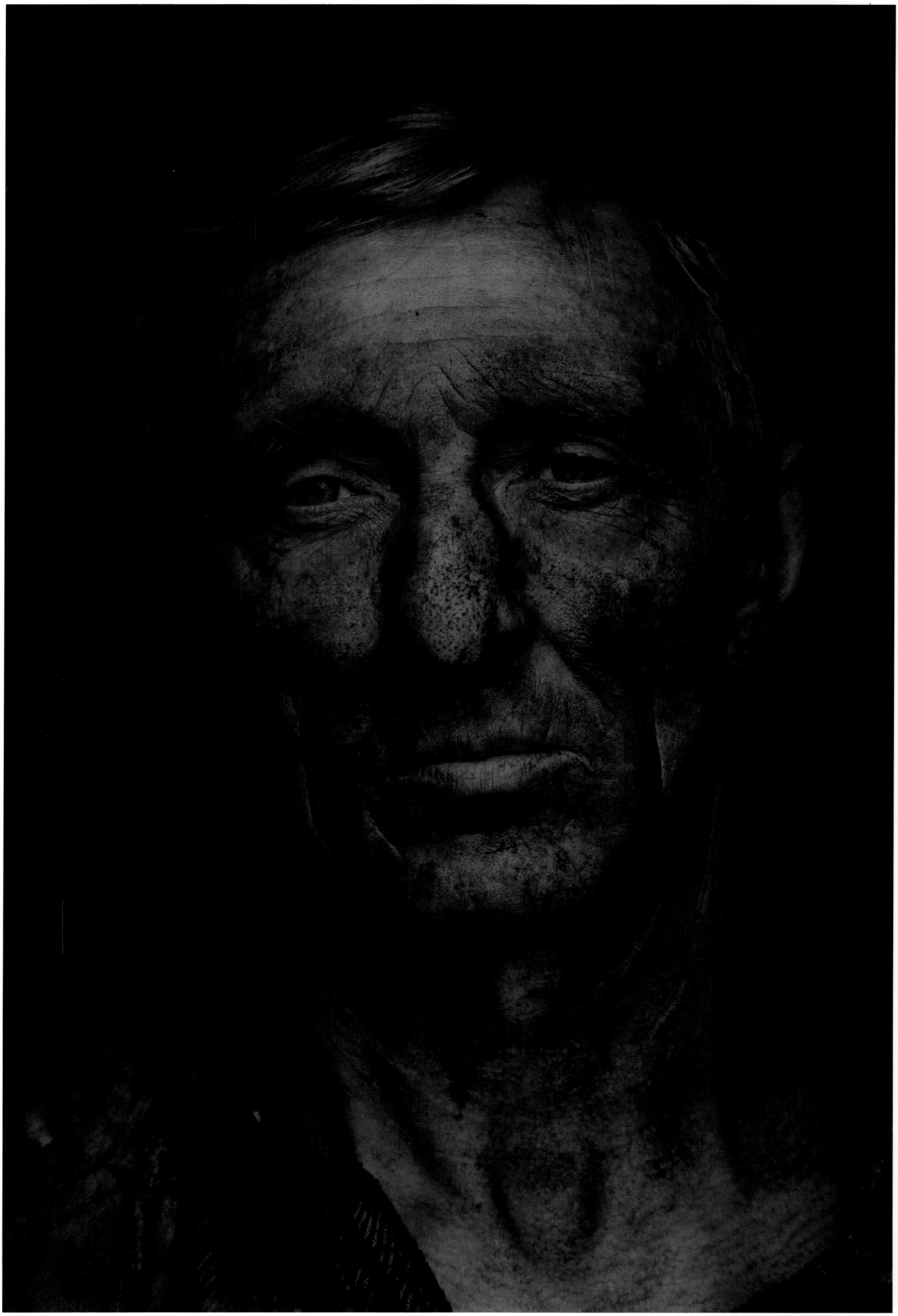

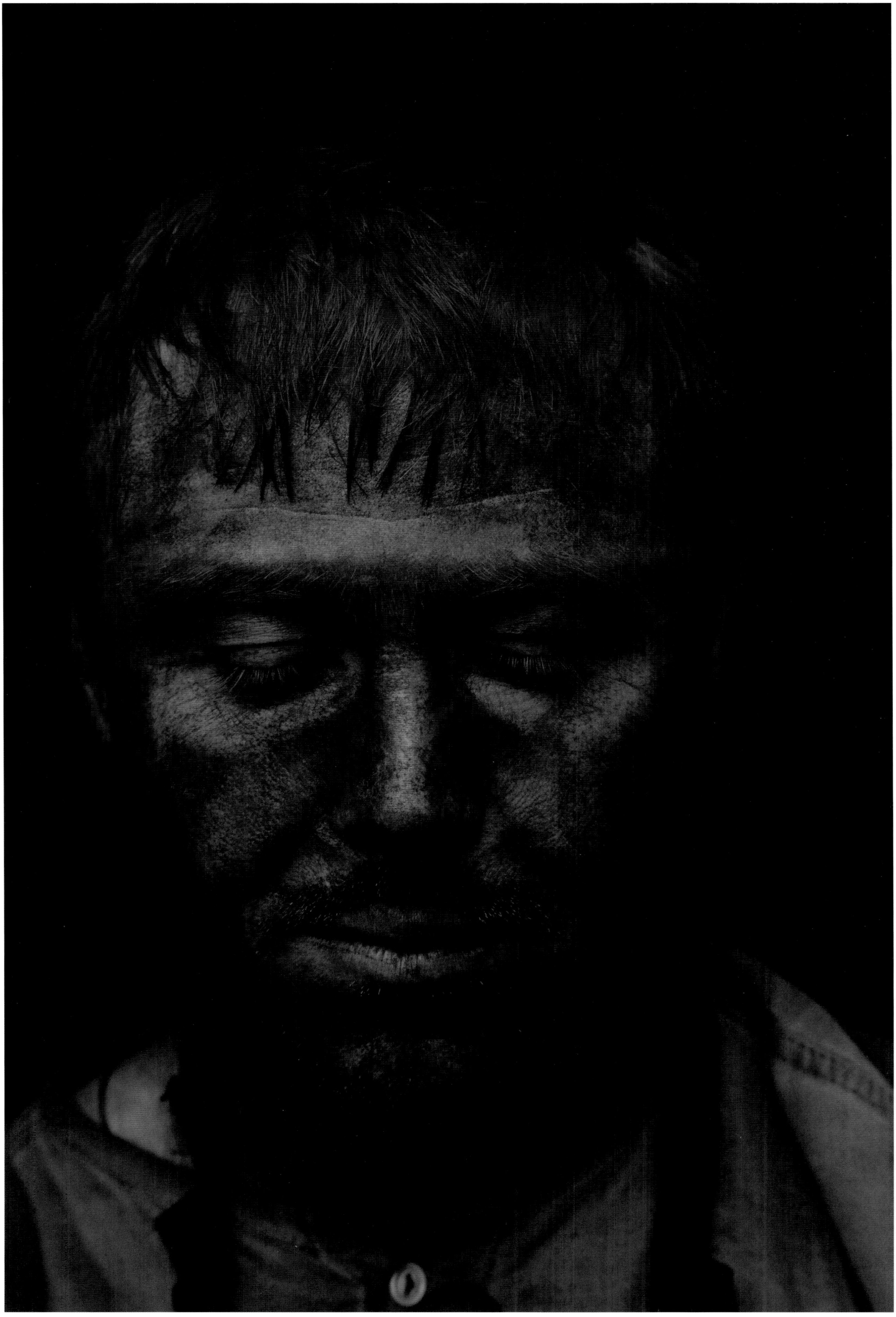

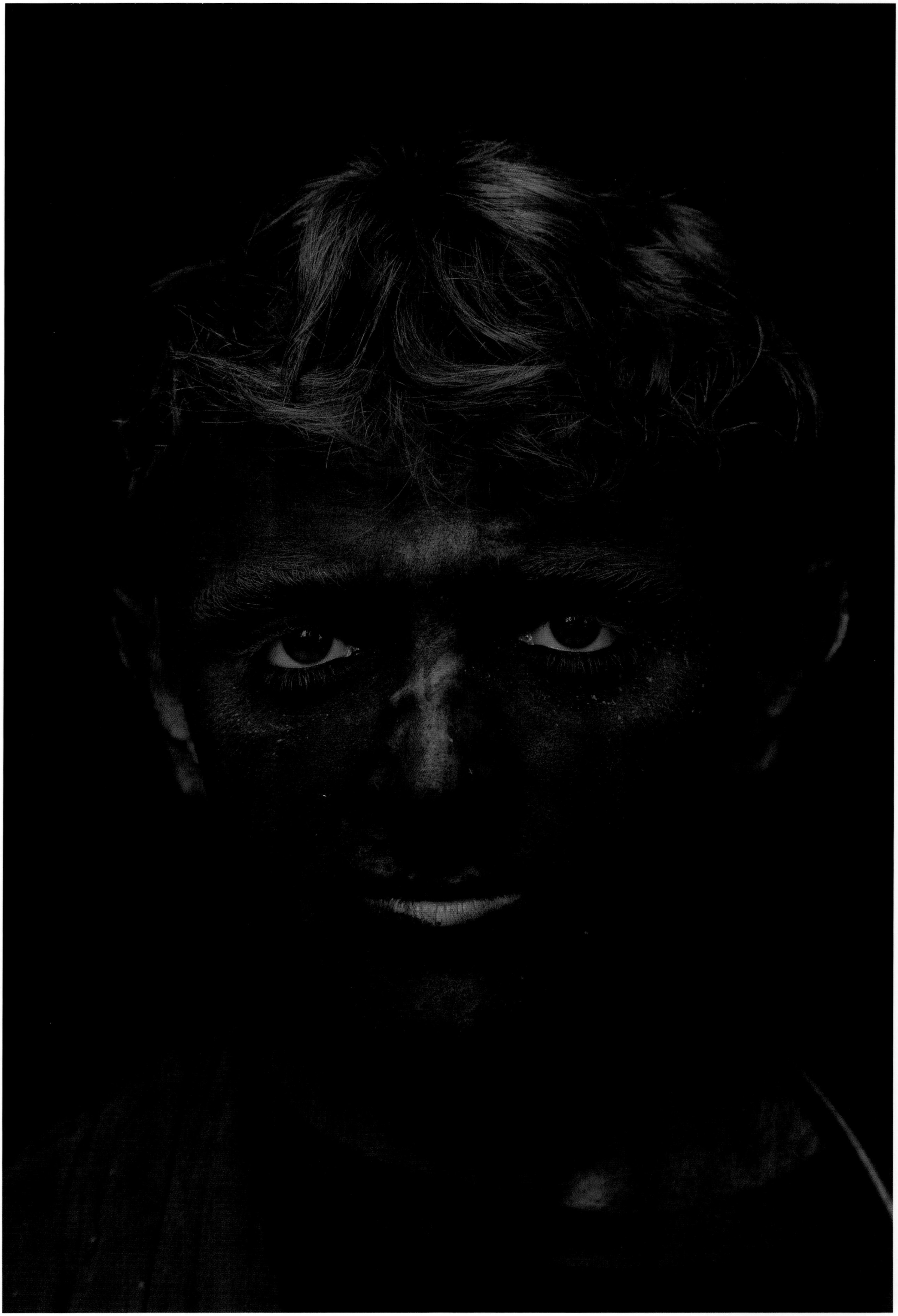

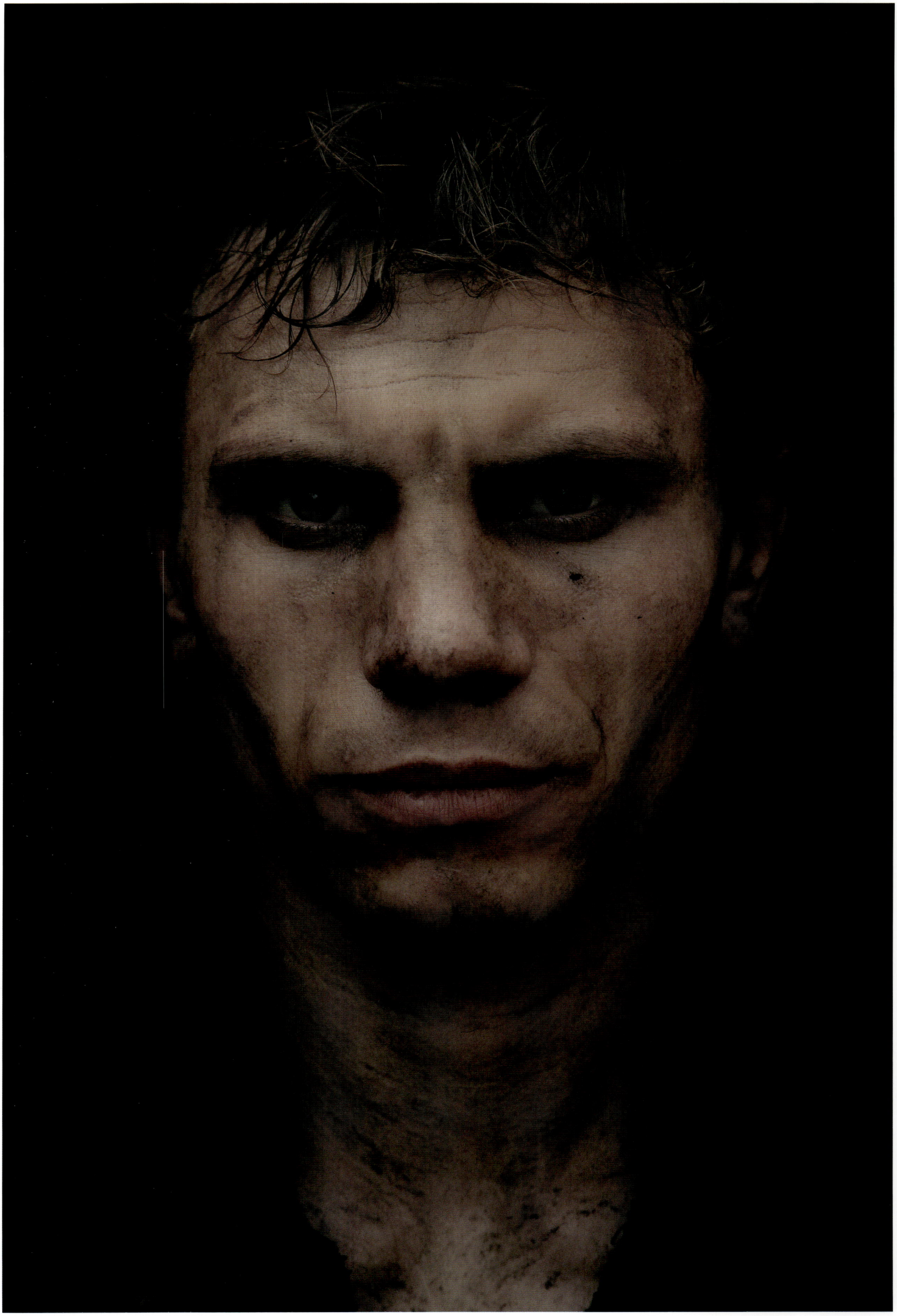

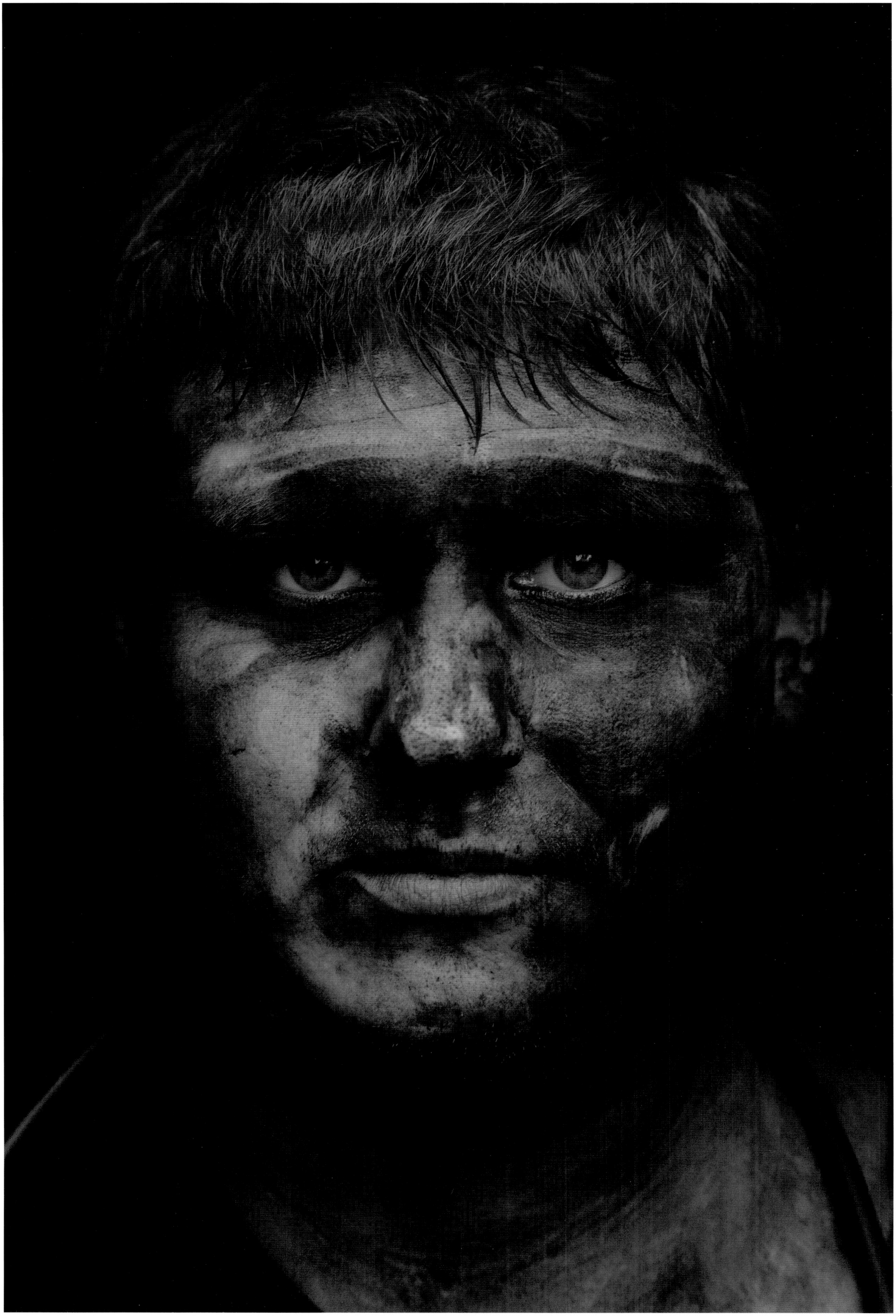

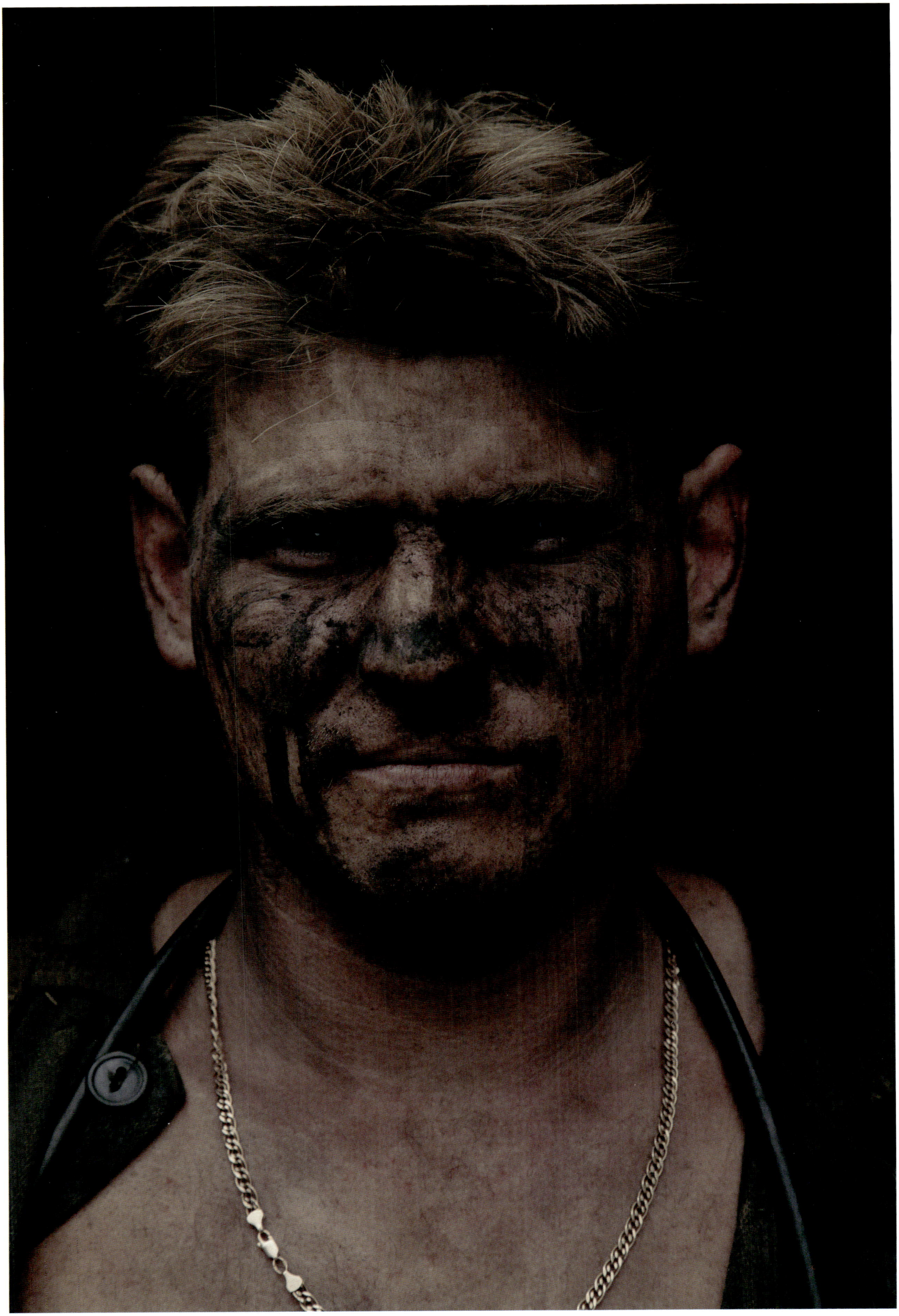

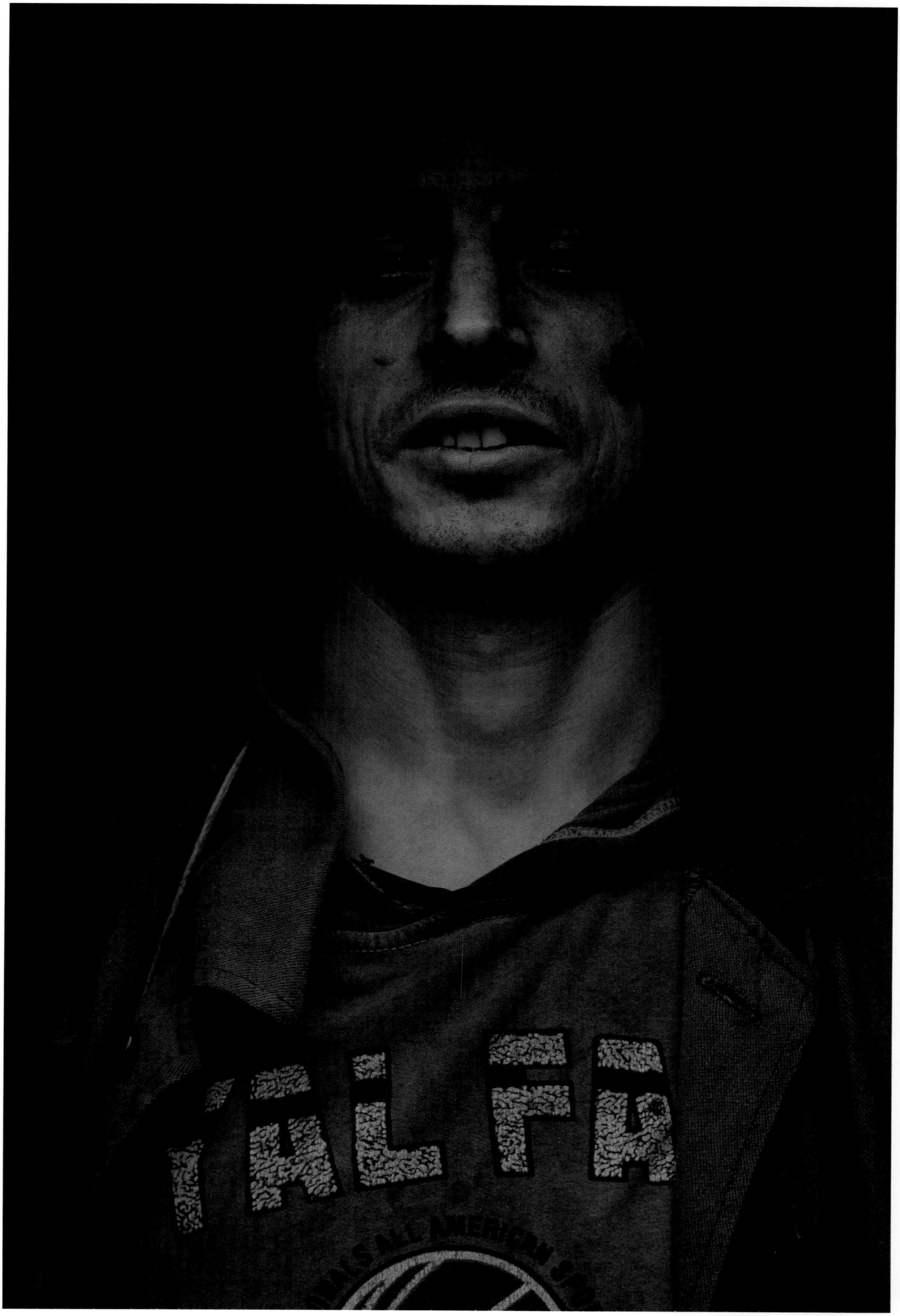
ALL AMERICAN

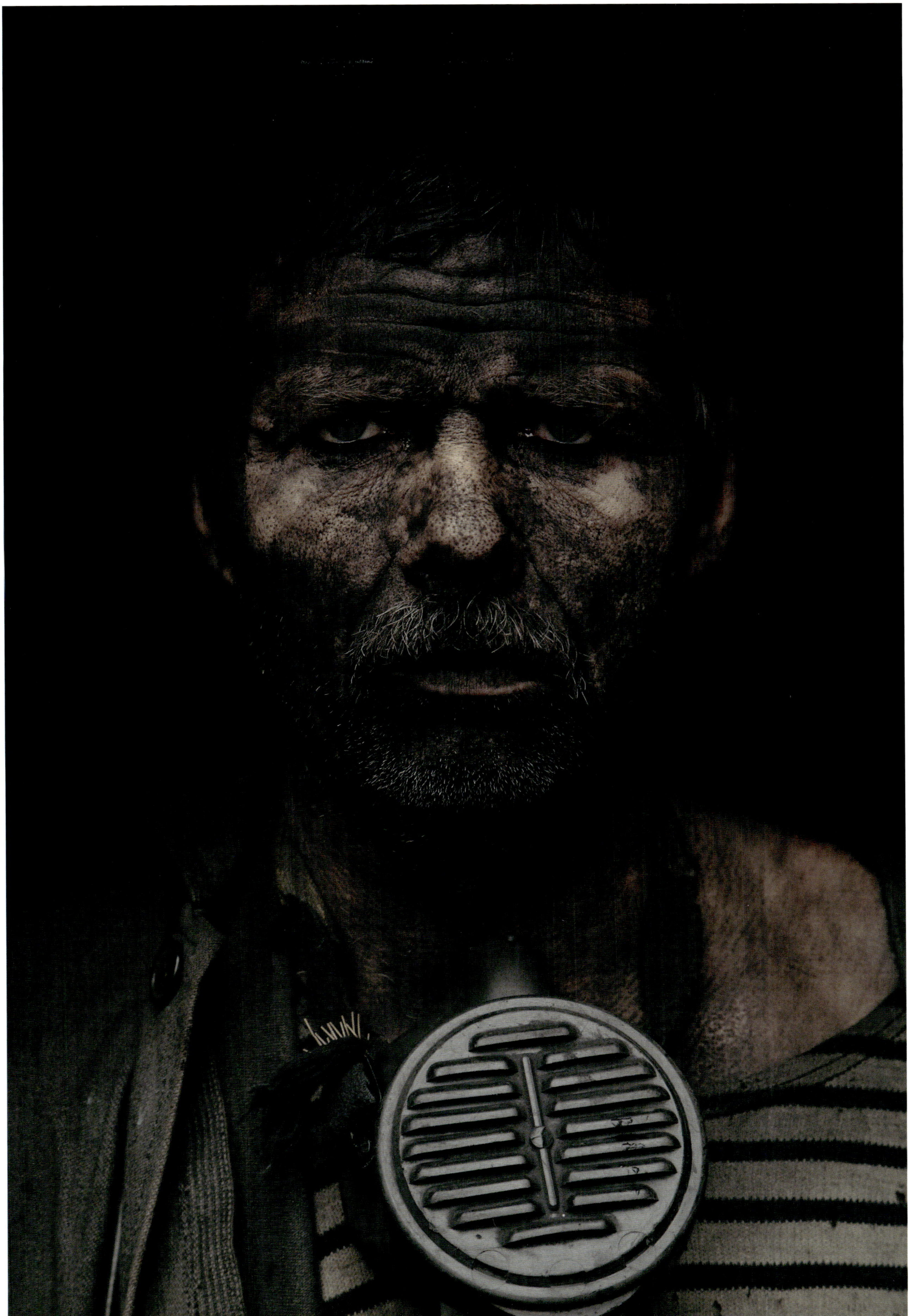

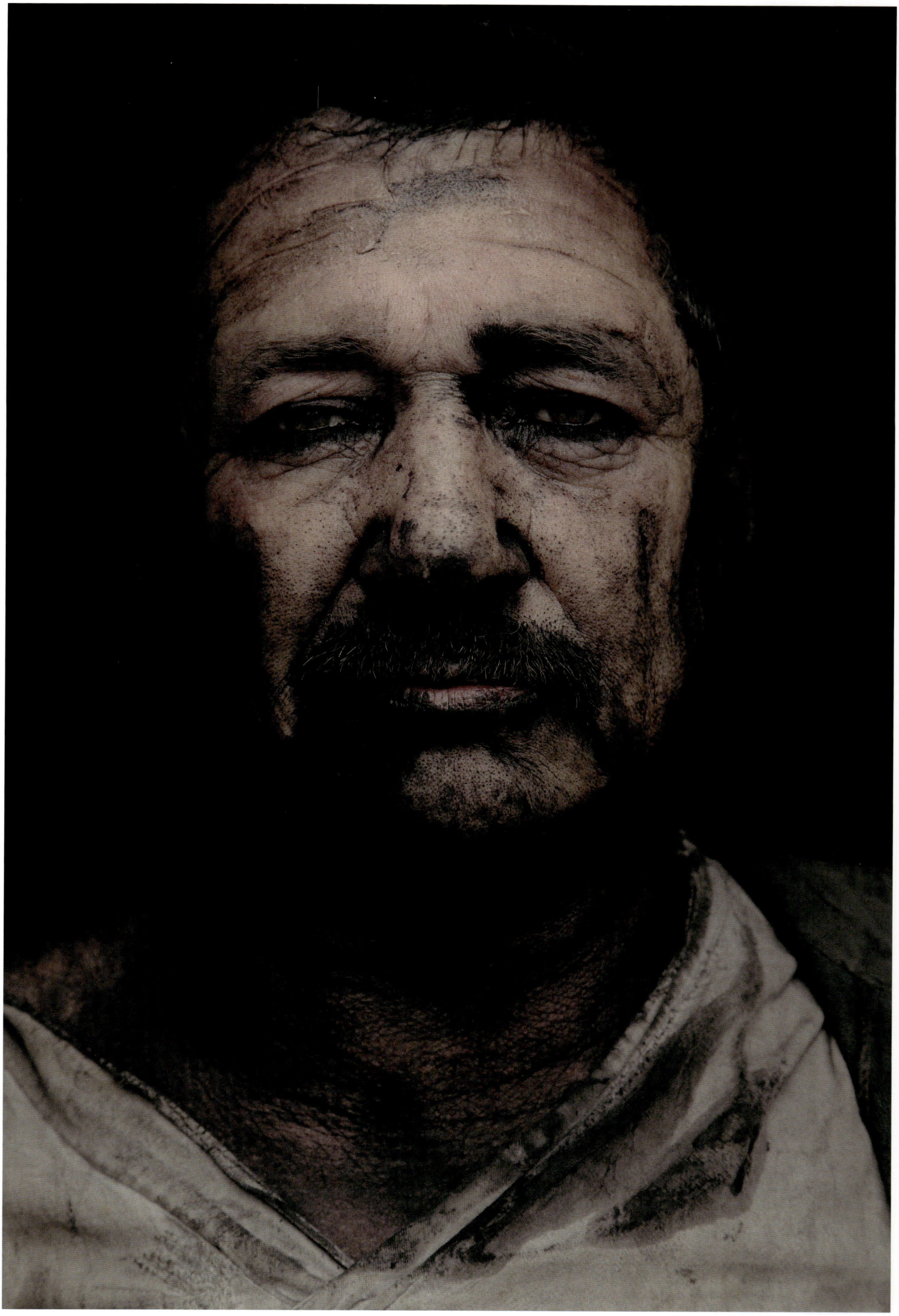

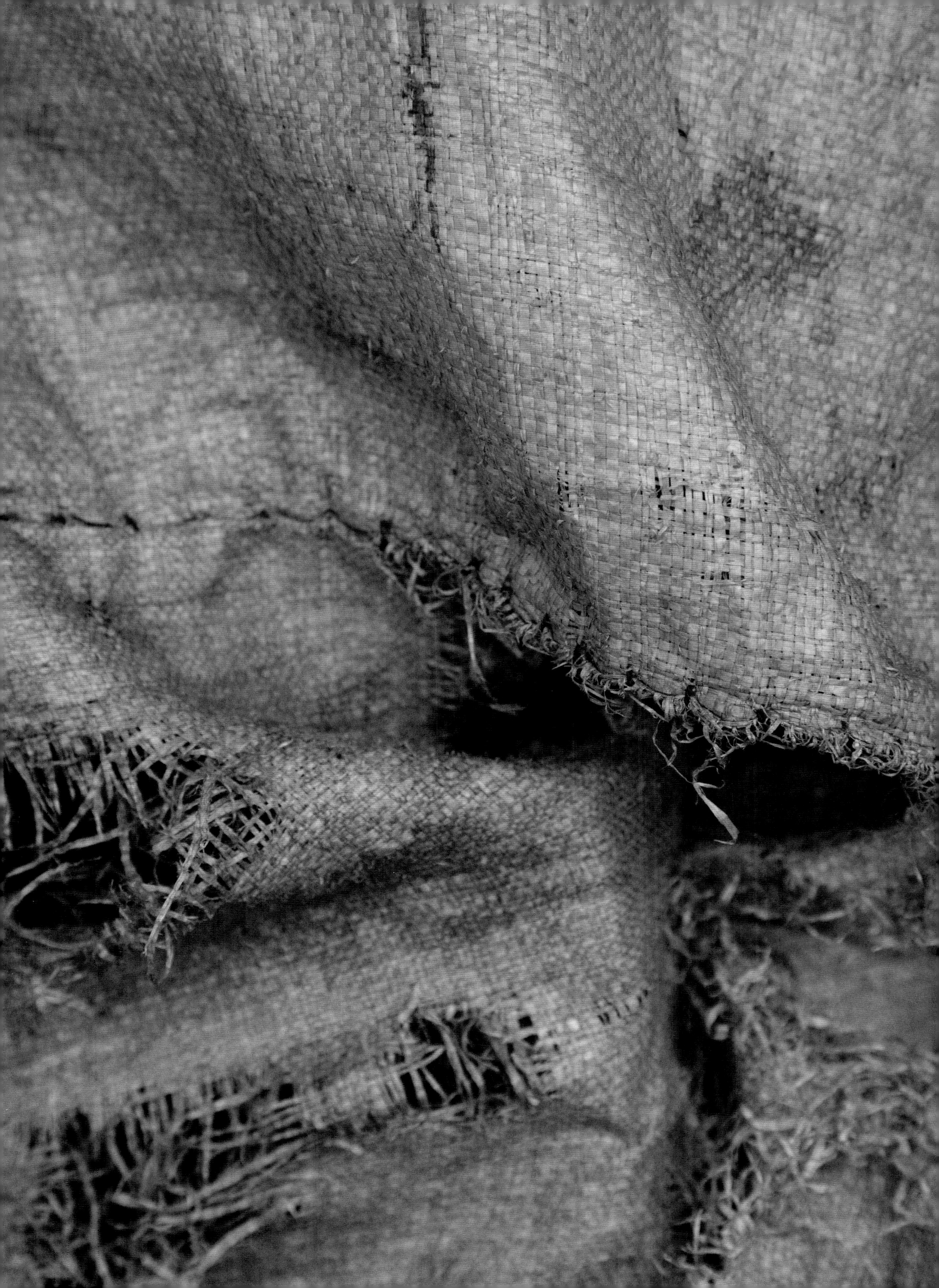

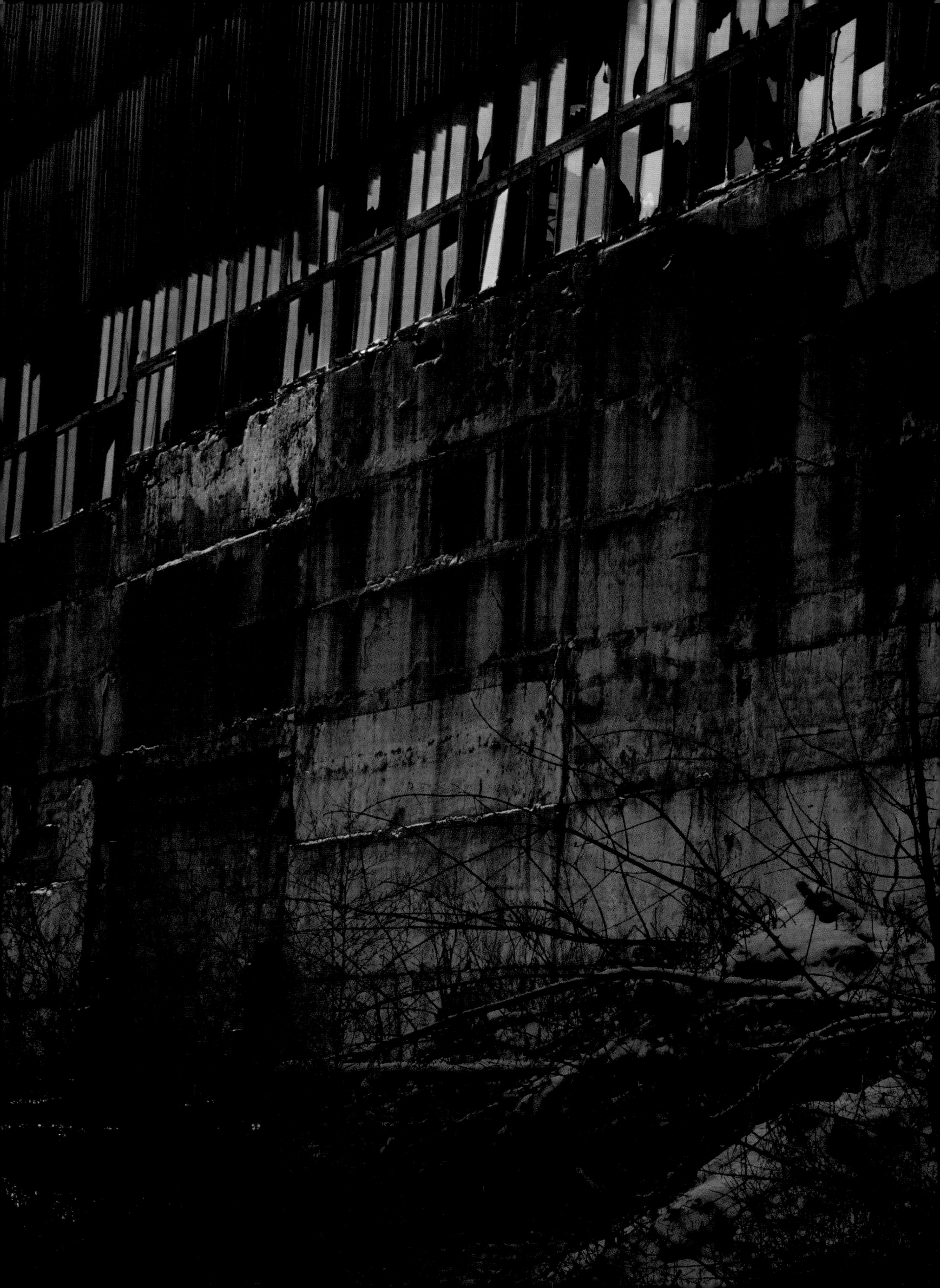

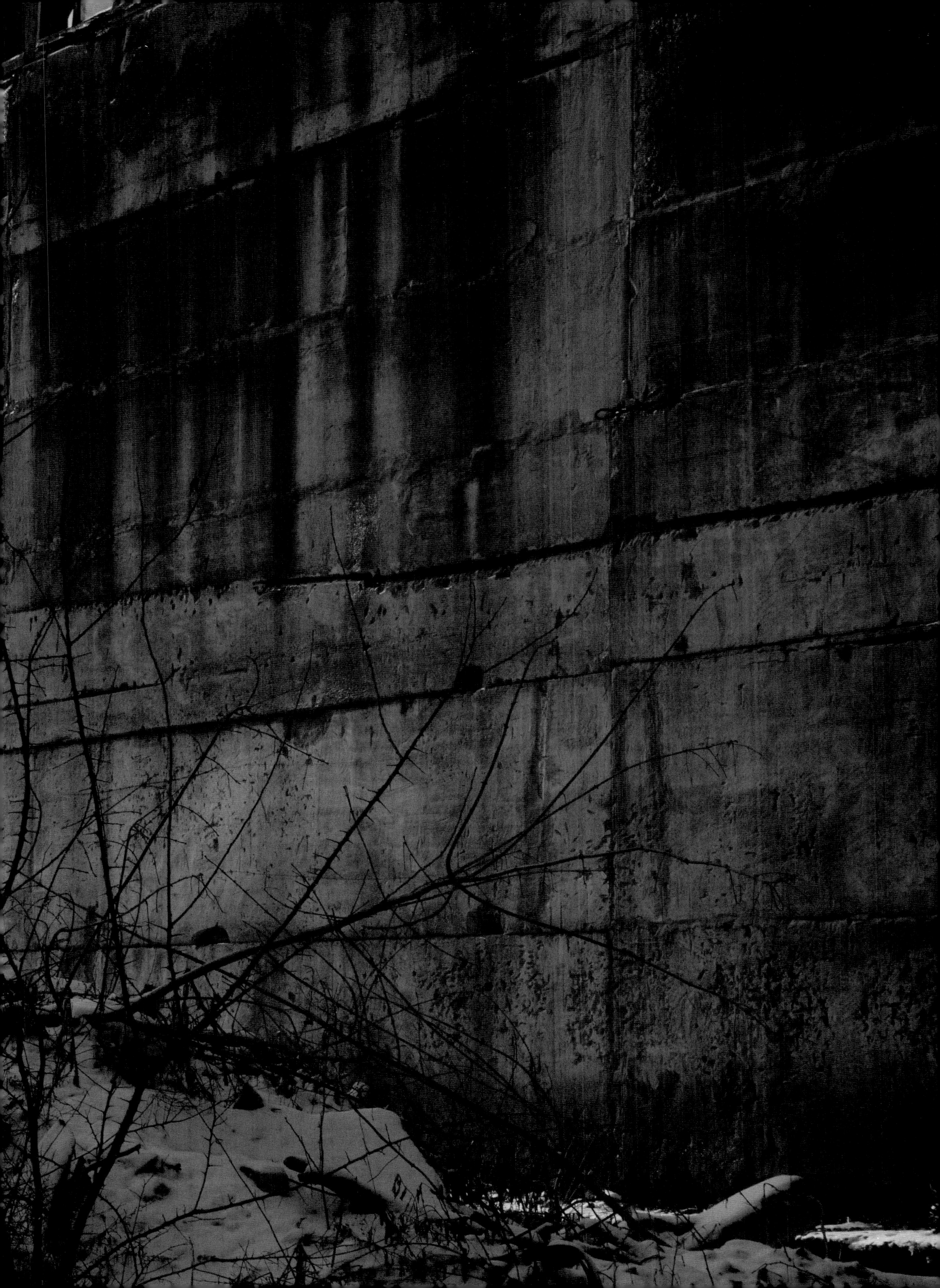

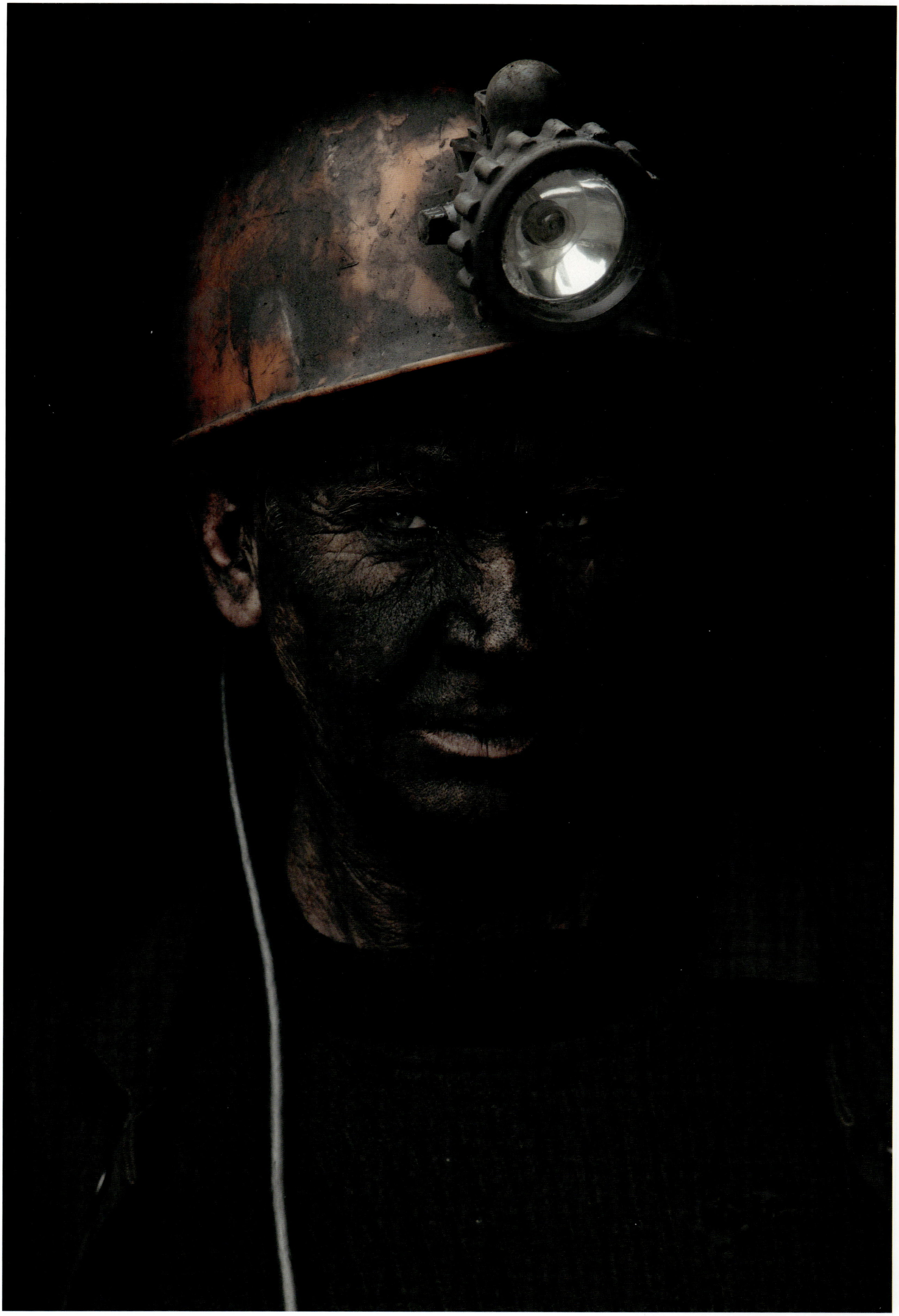

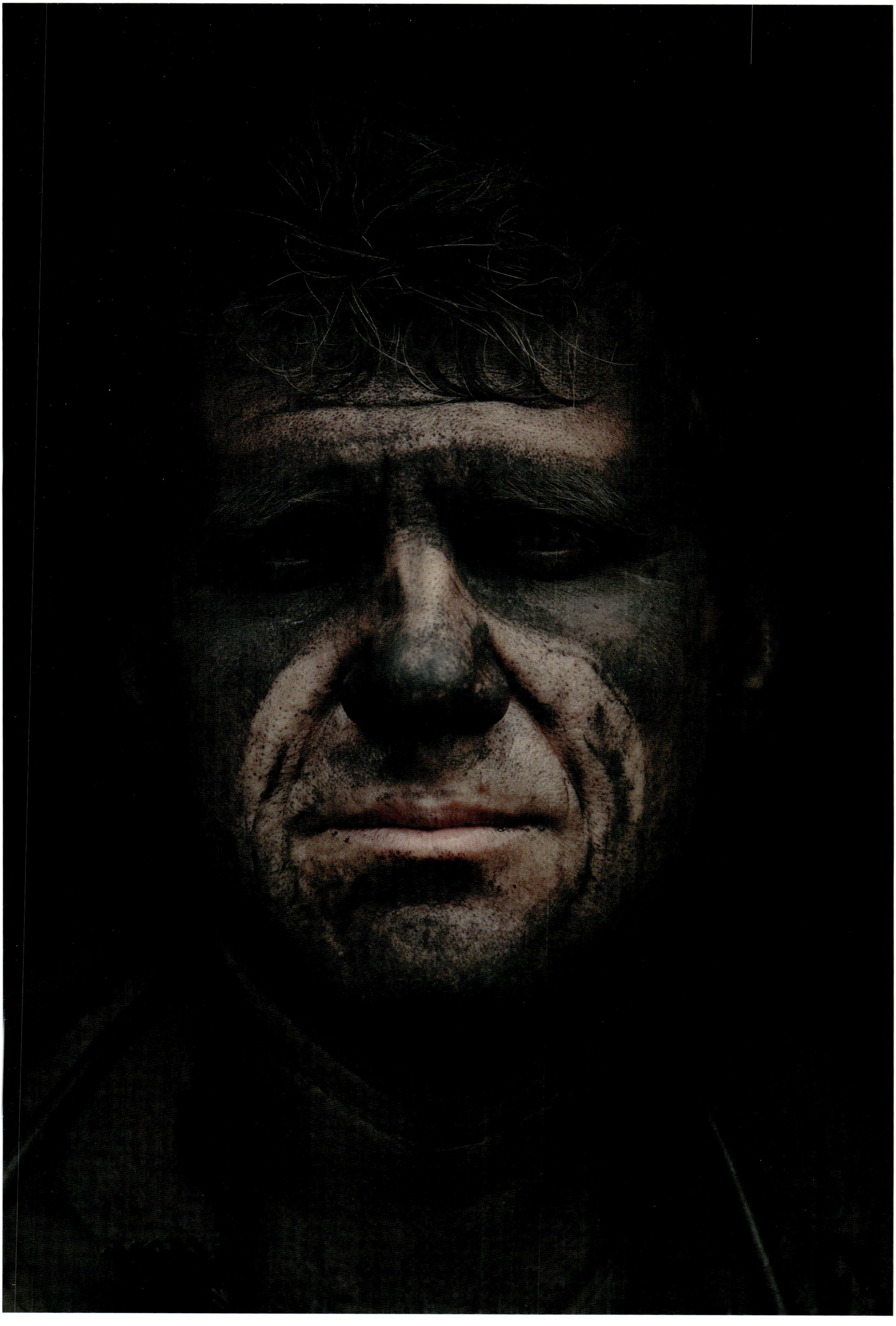

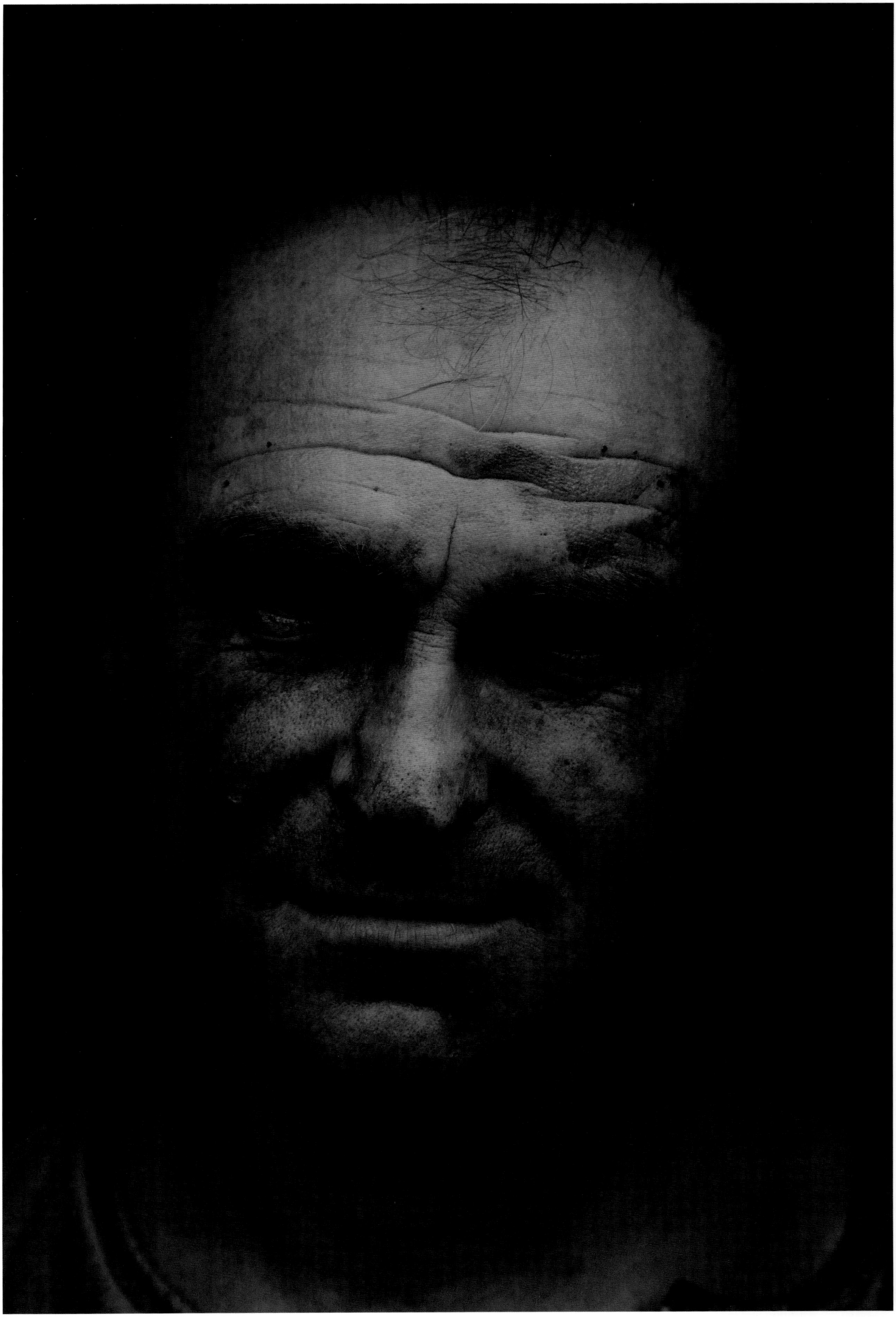

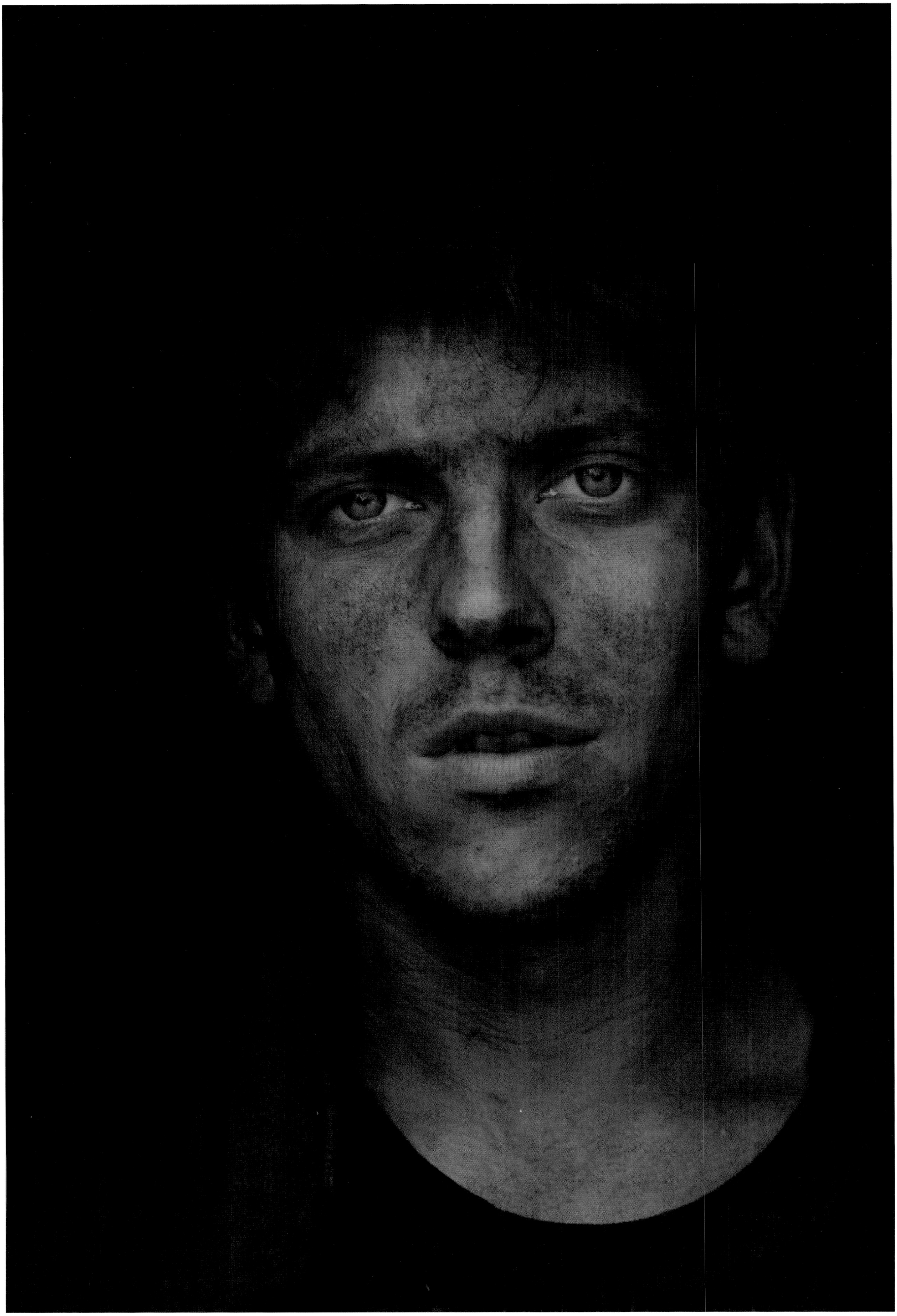

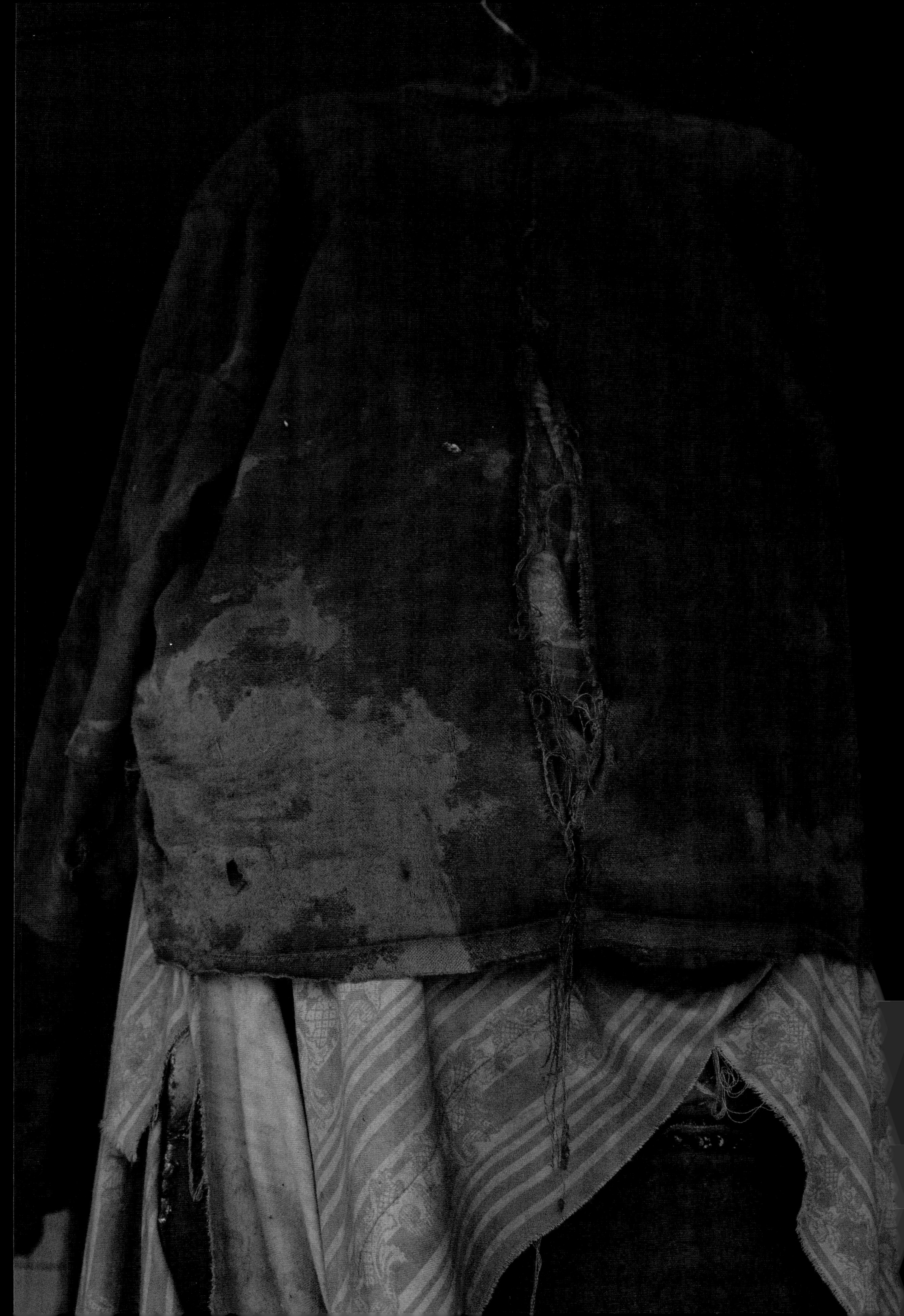

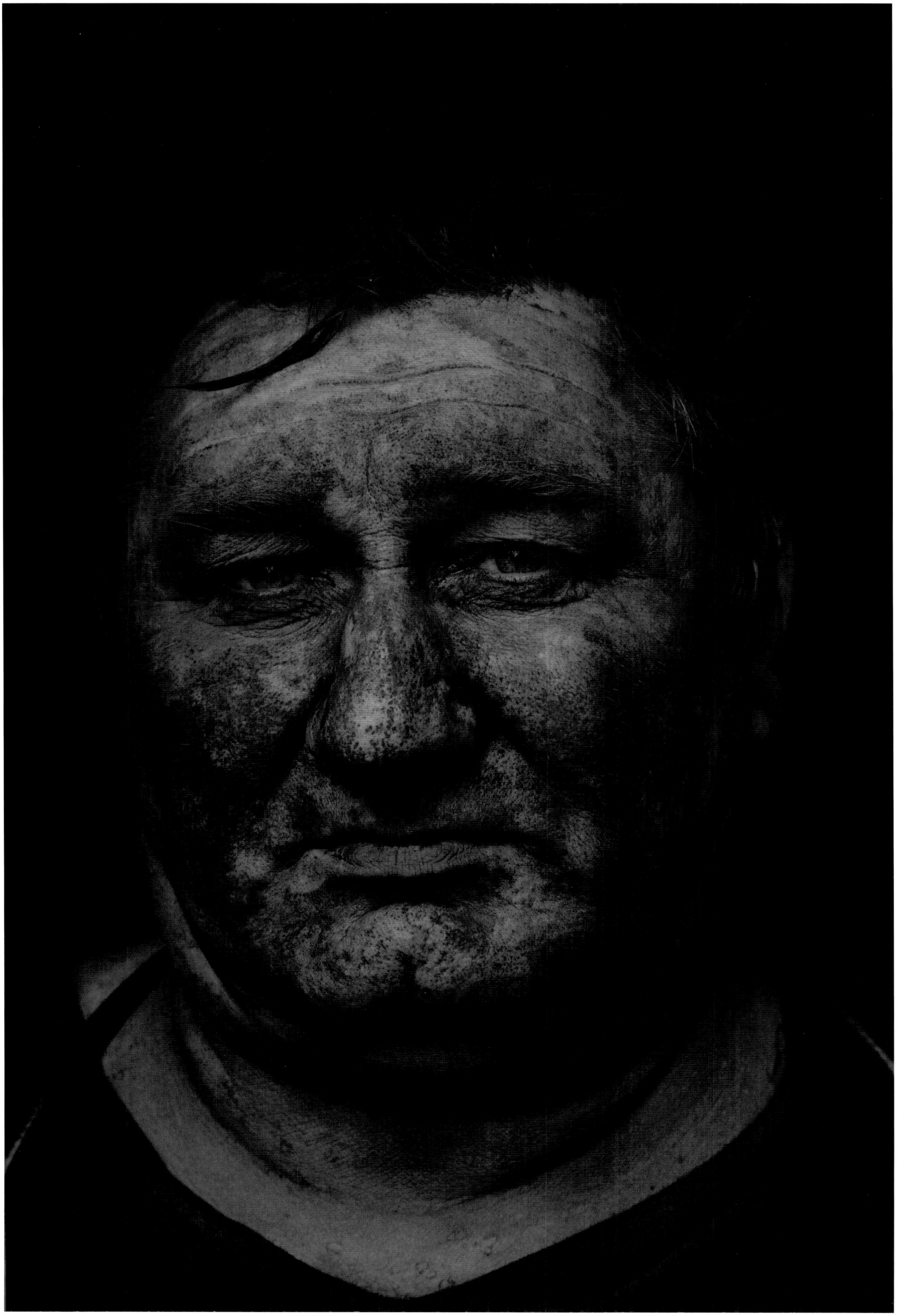

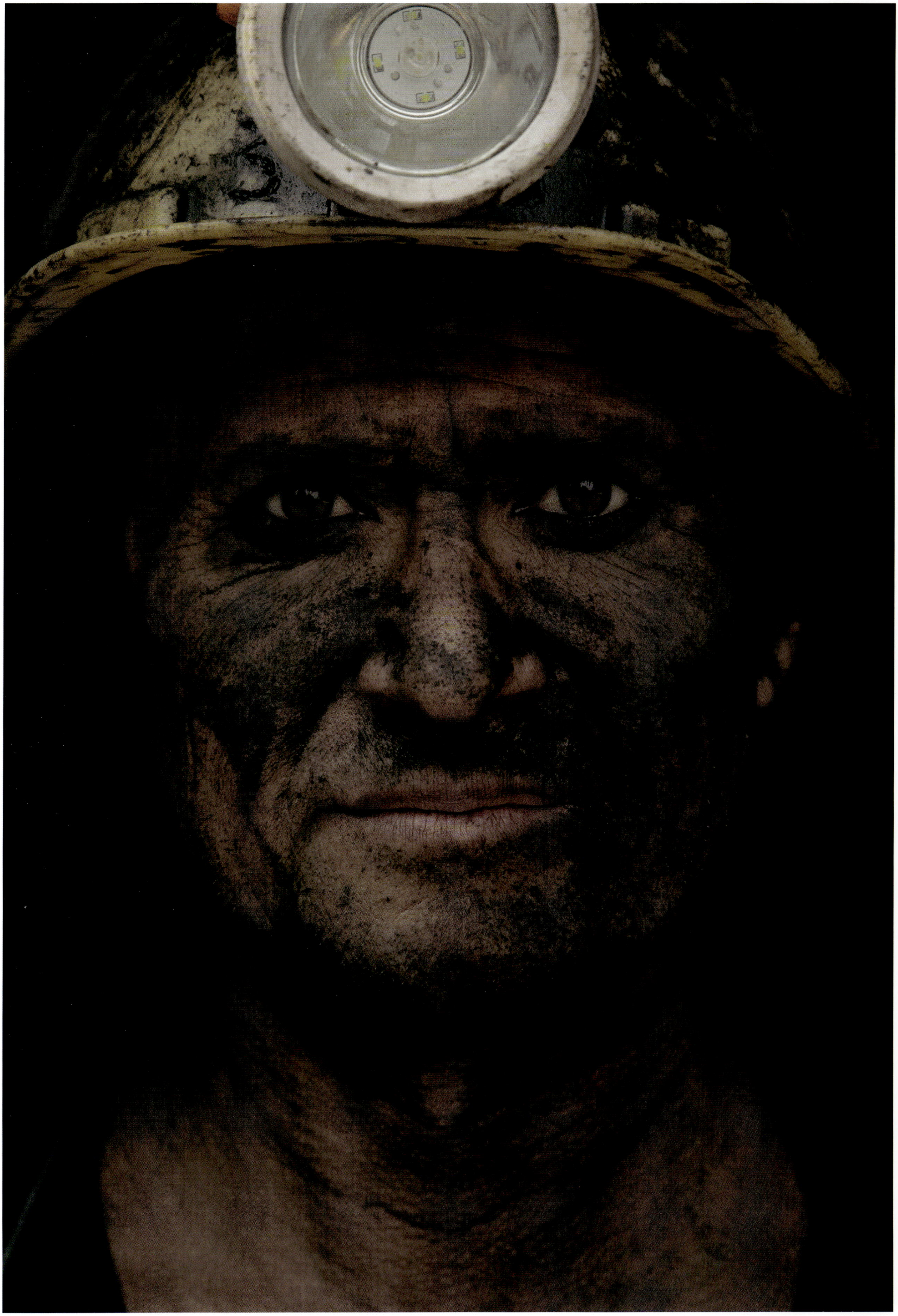

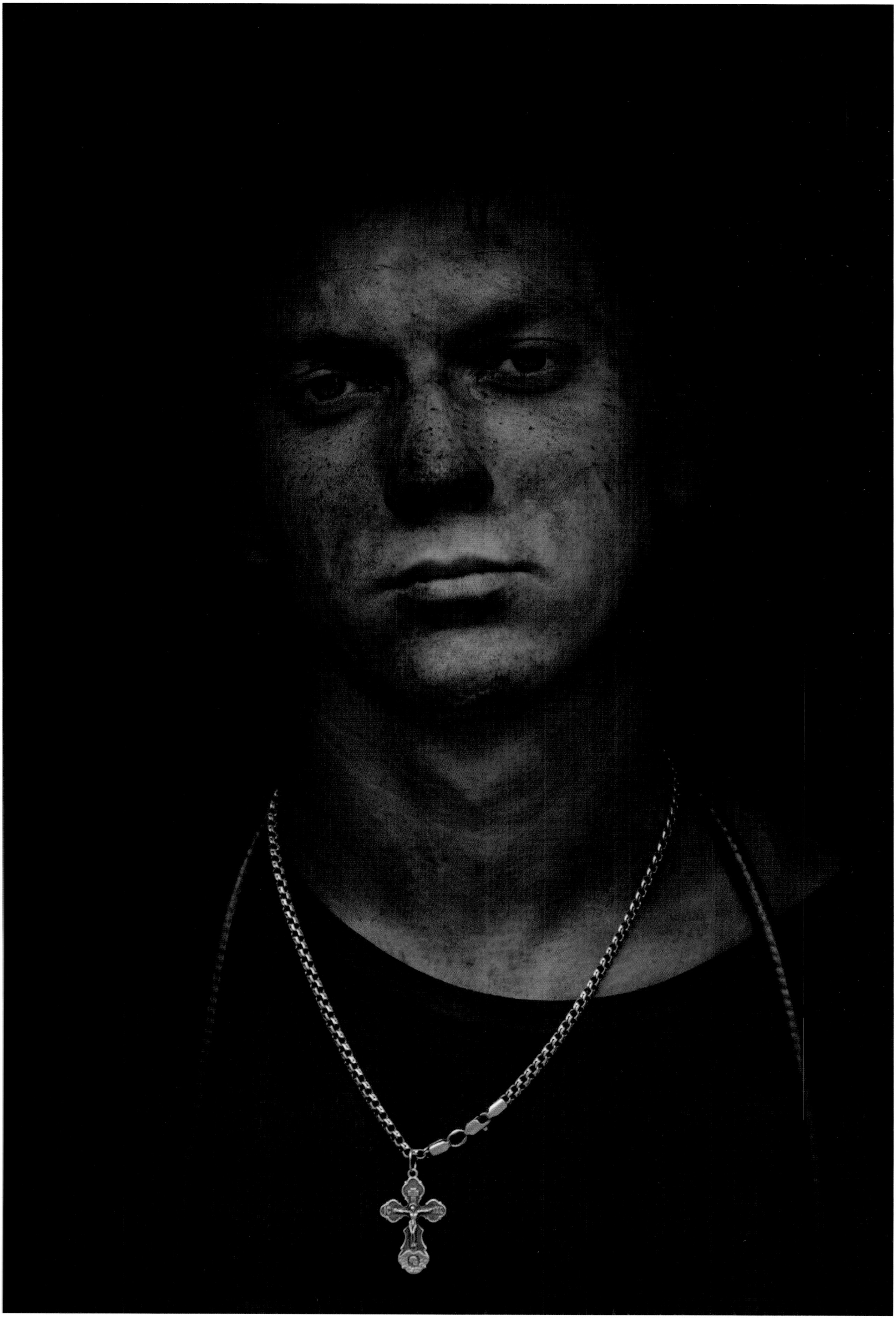

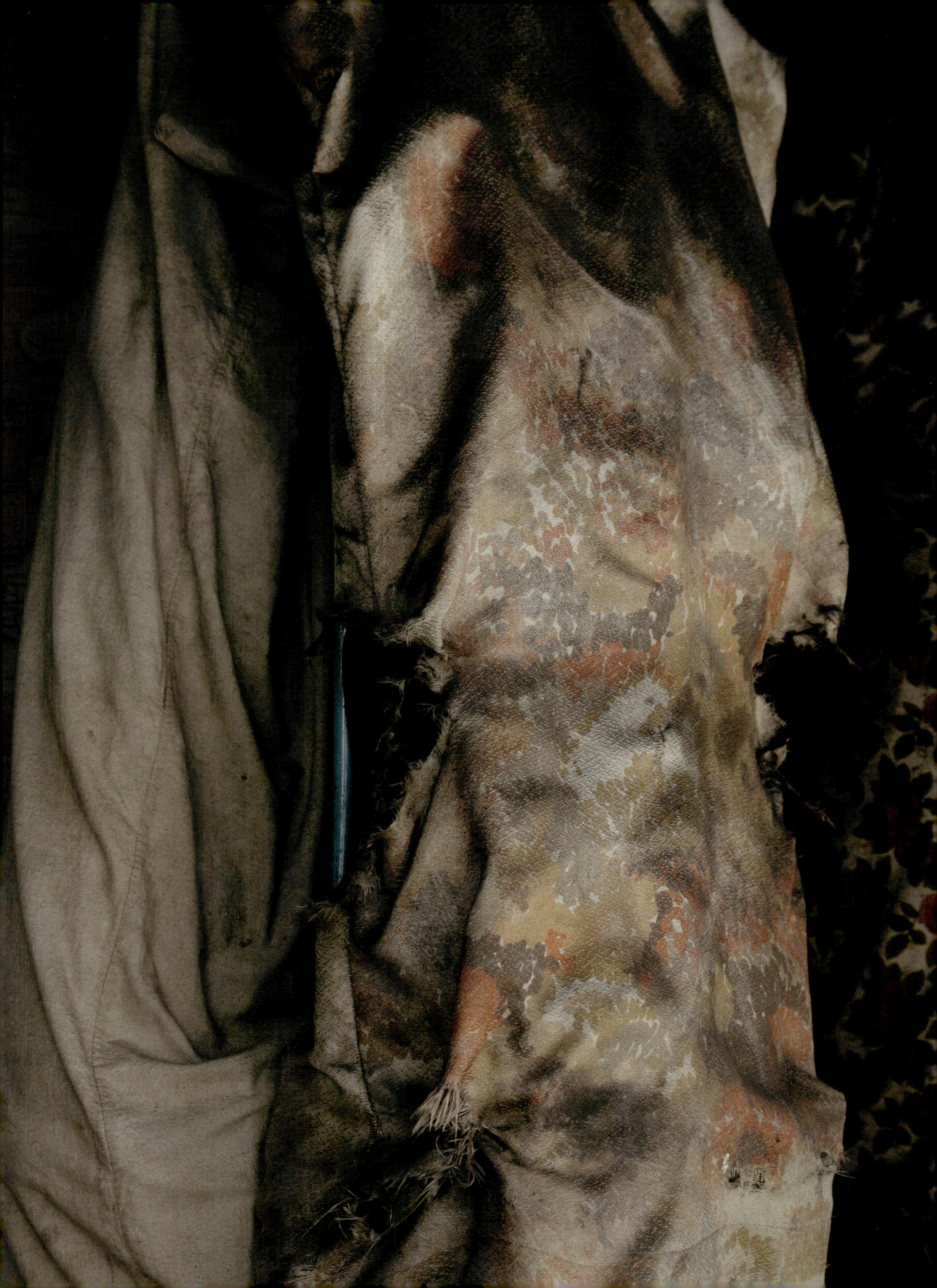

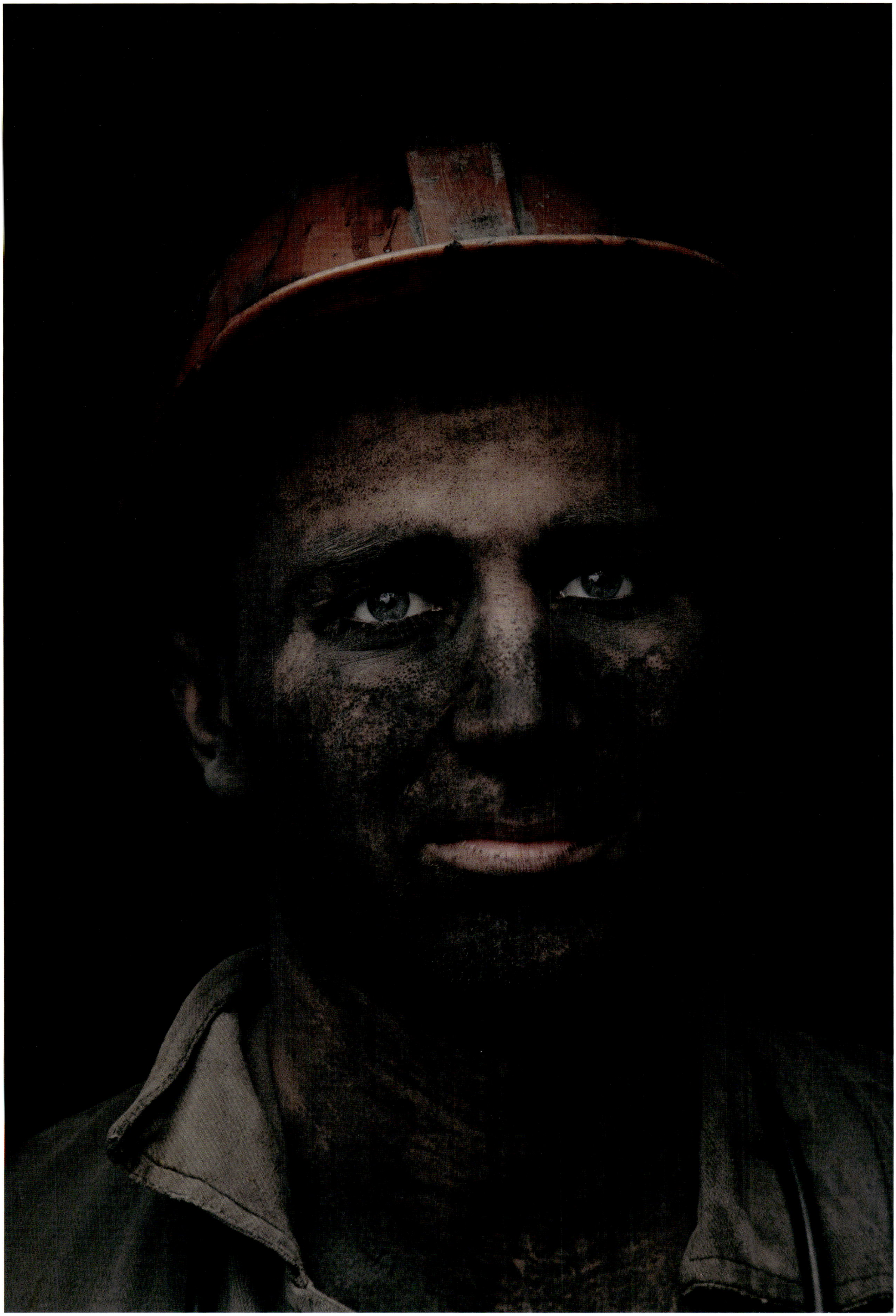

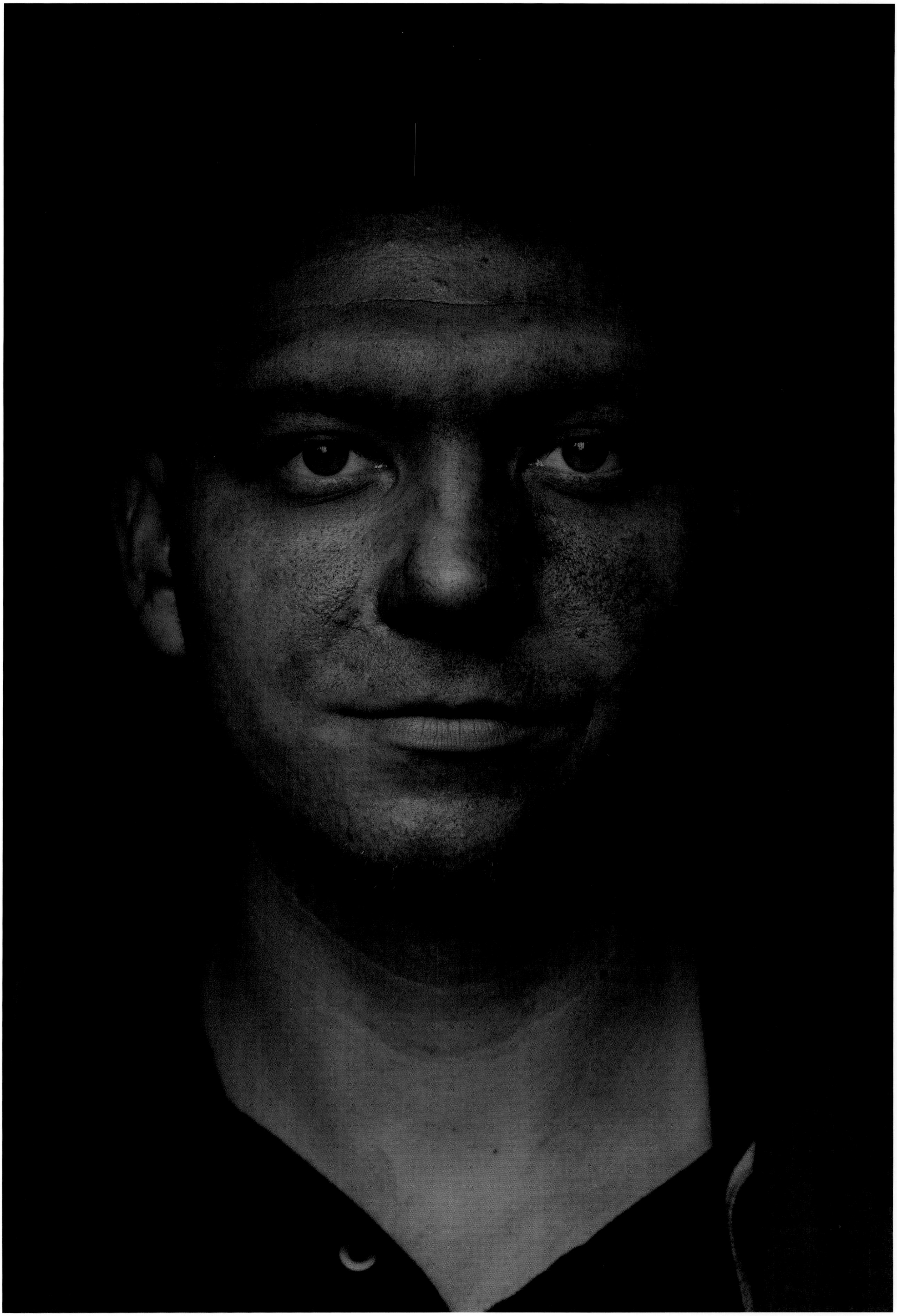

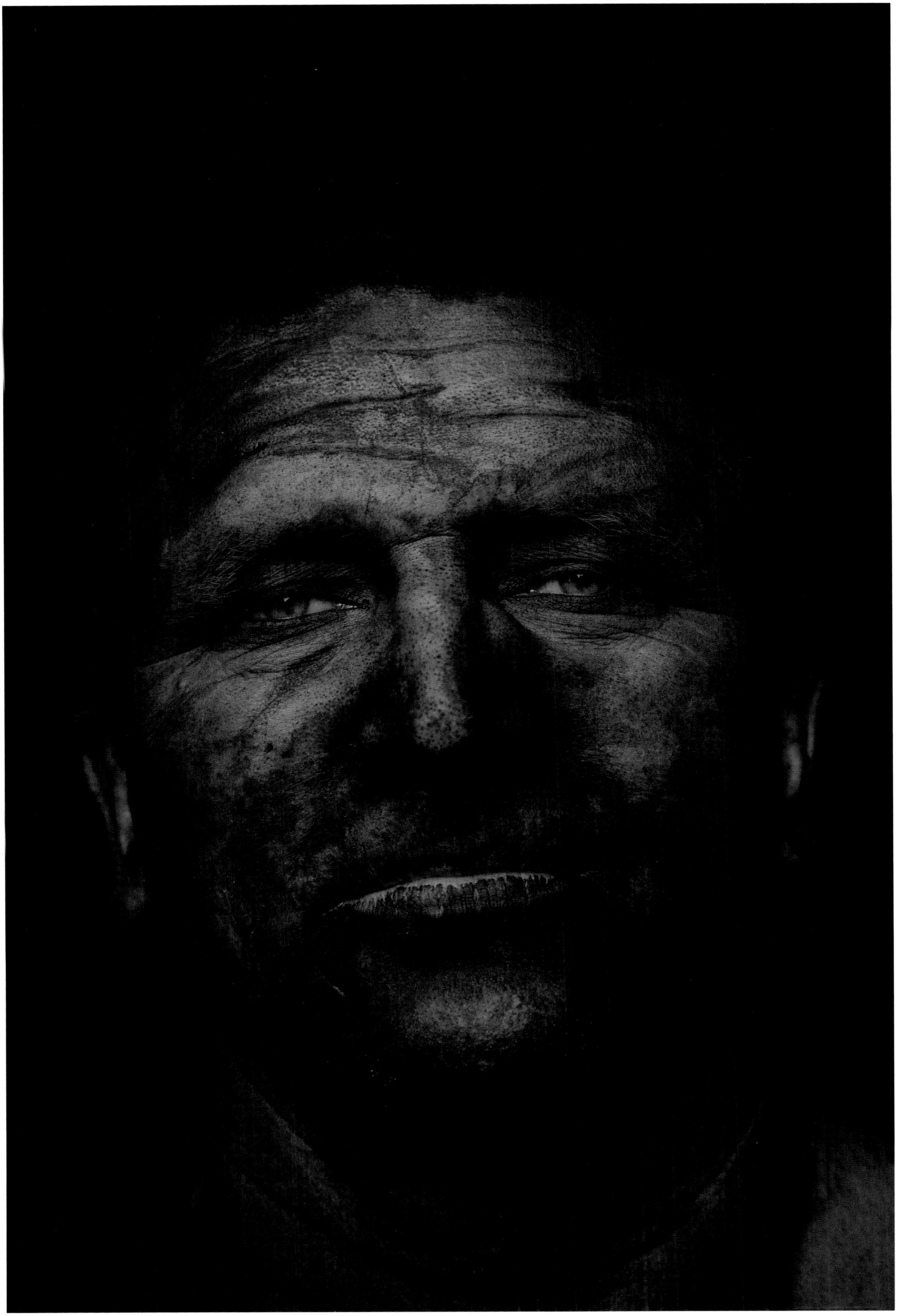

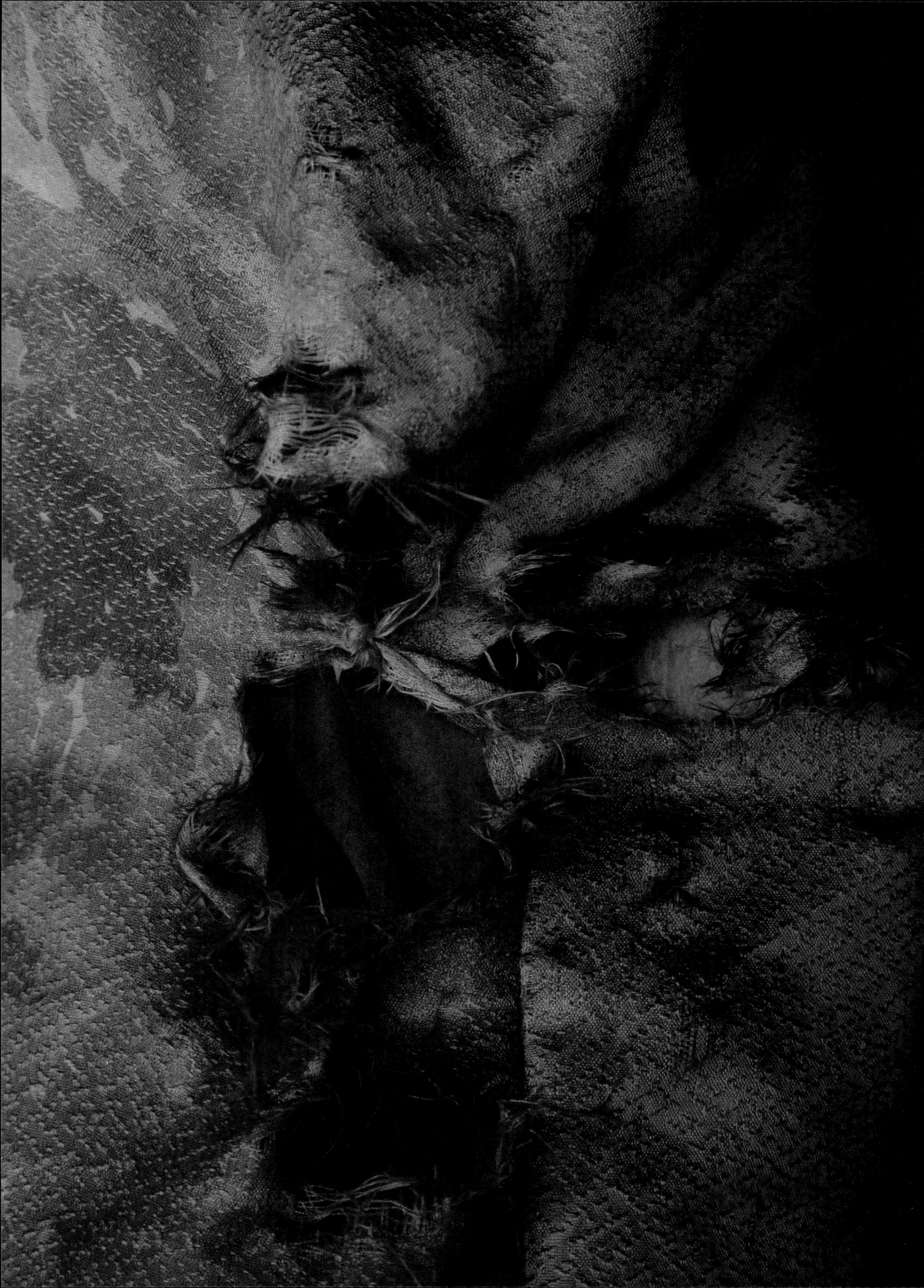

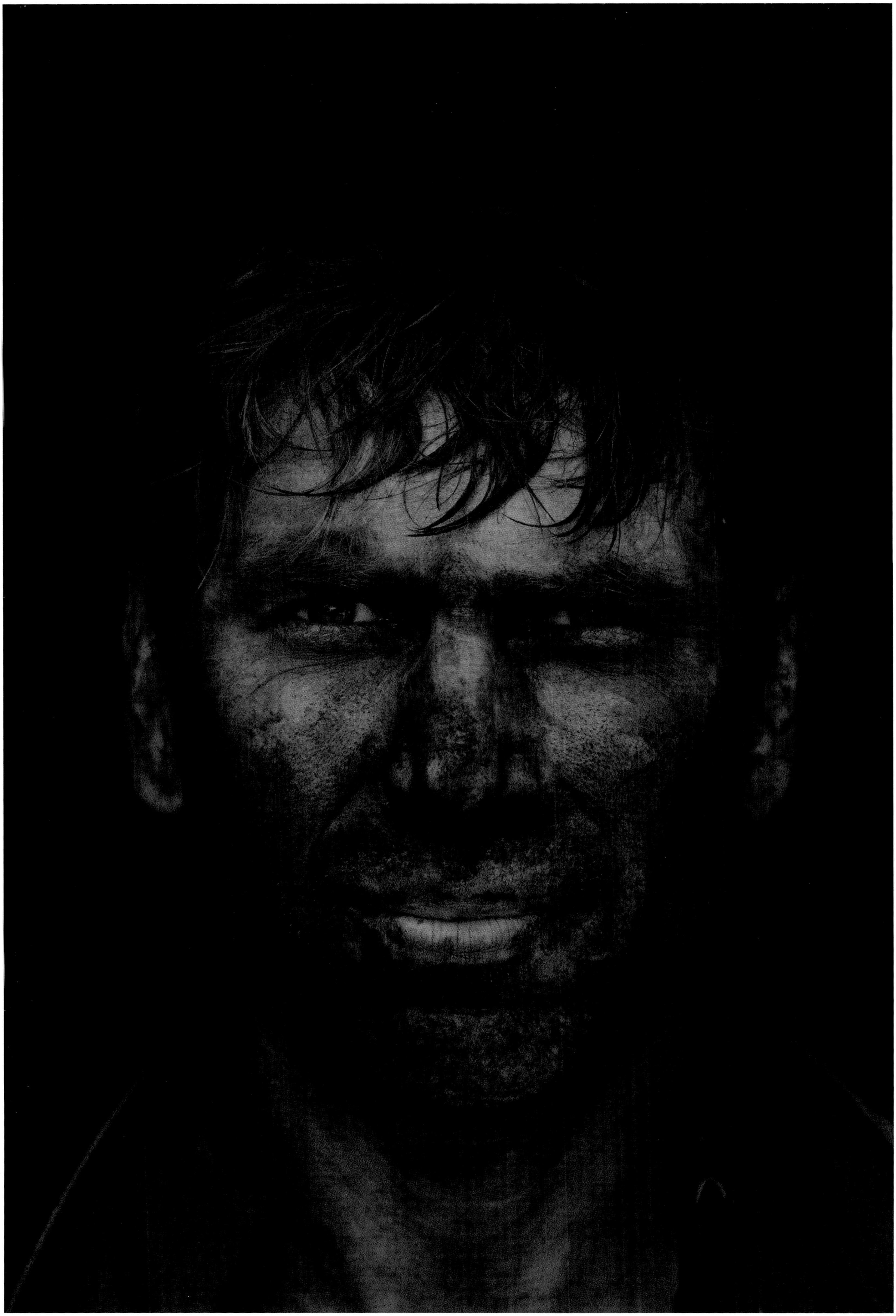

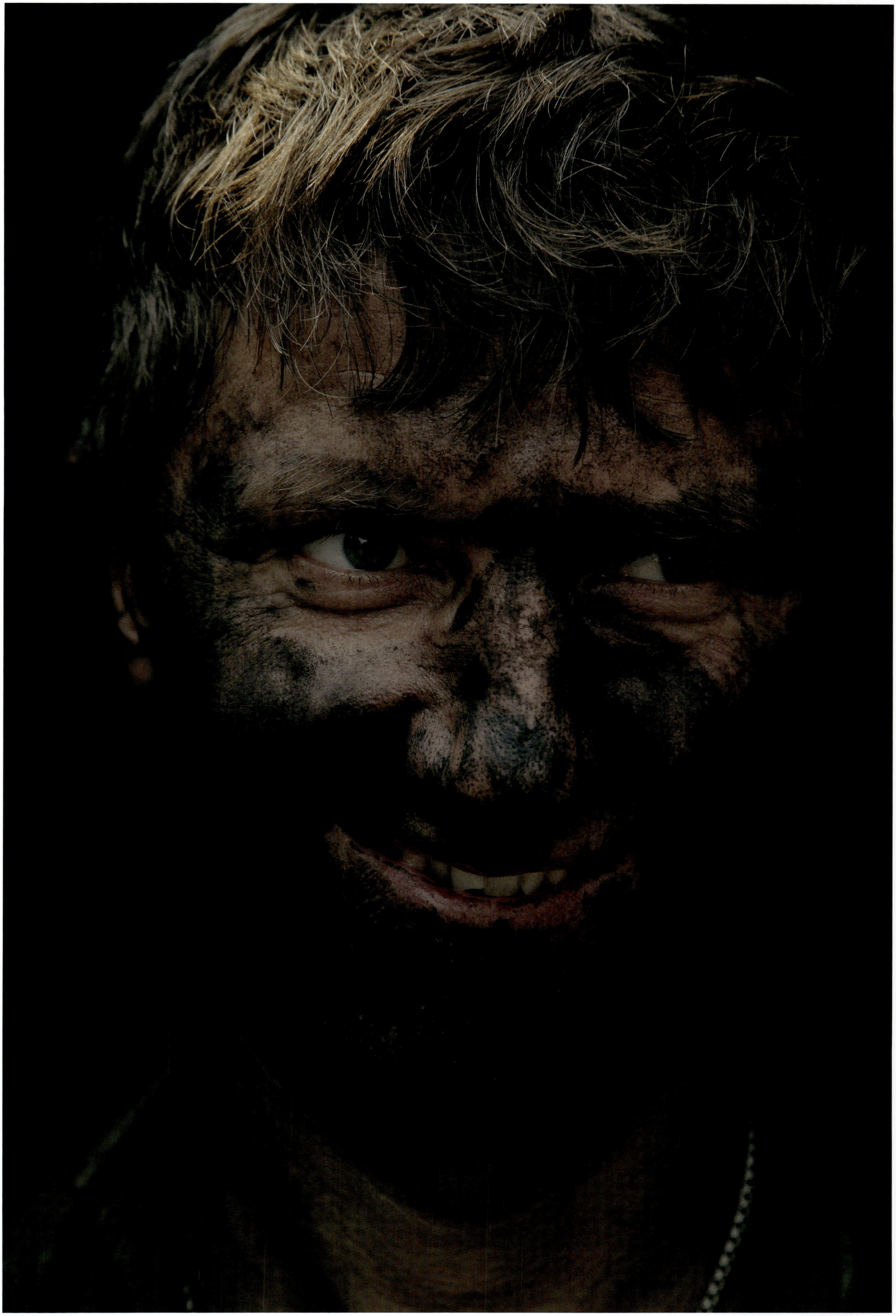

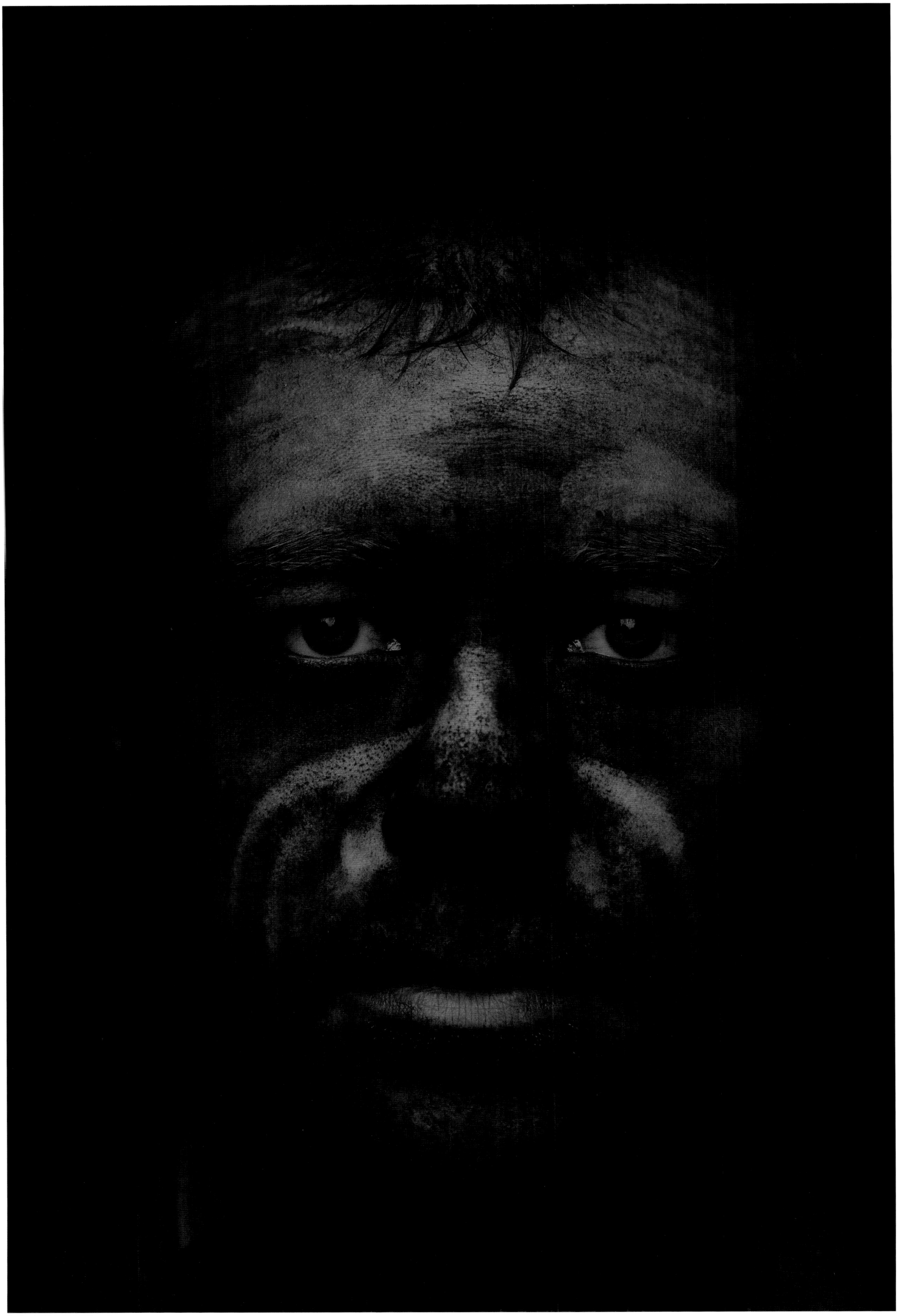

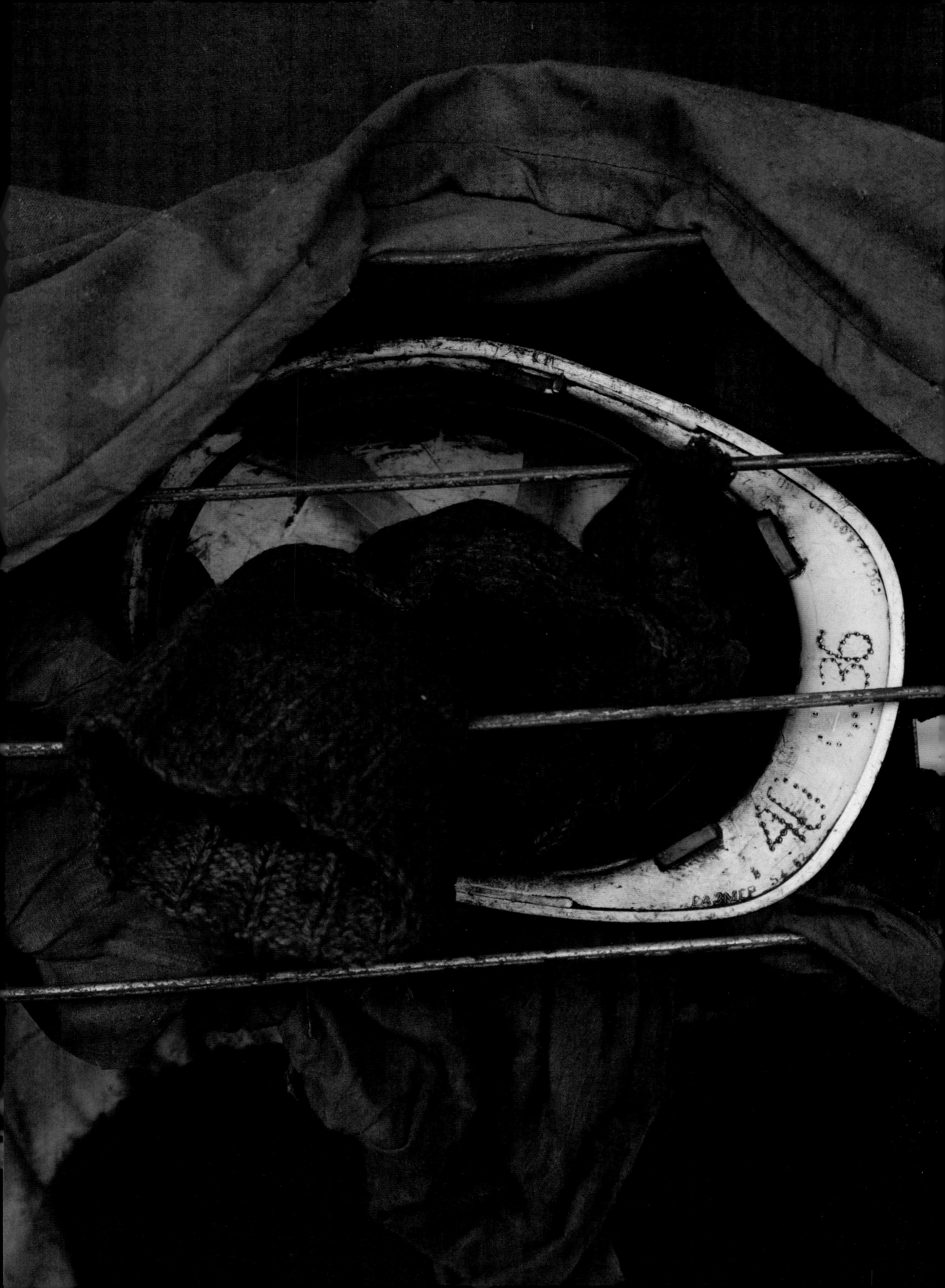
РАЗМЕР

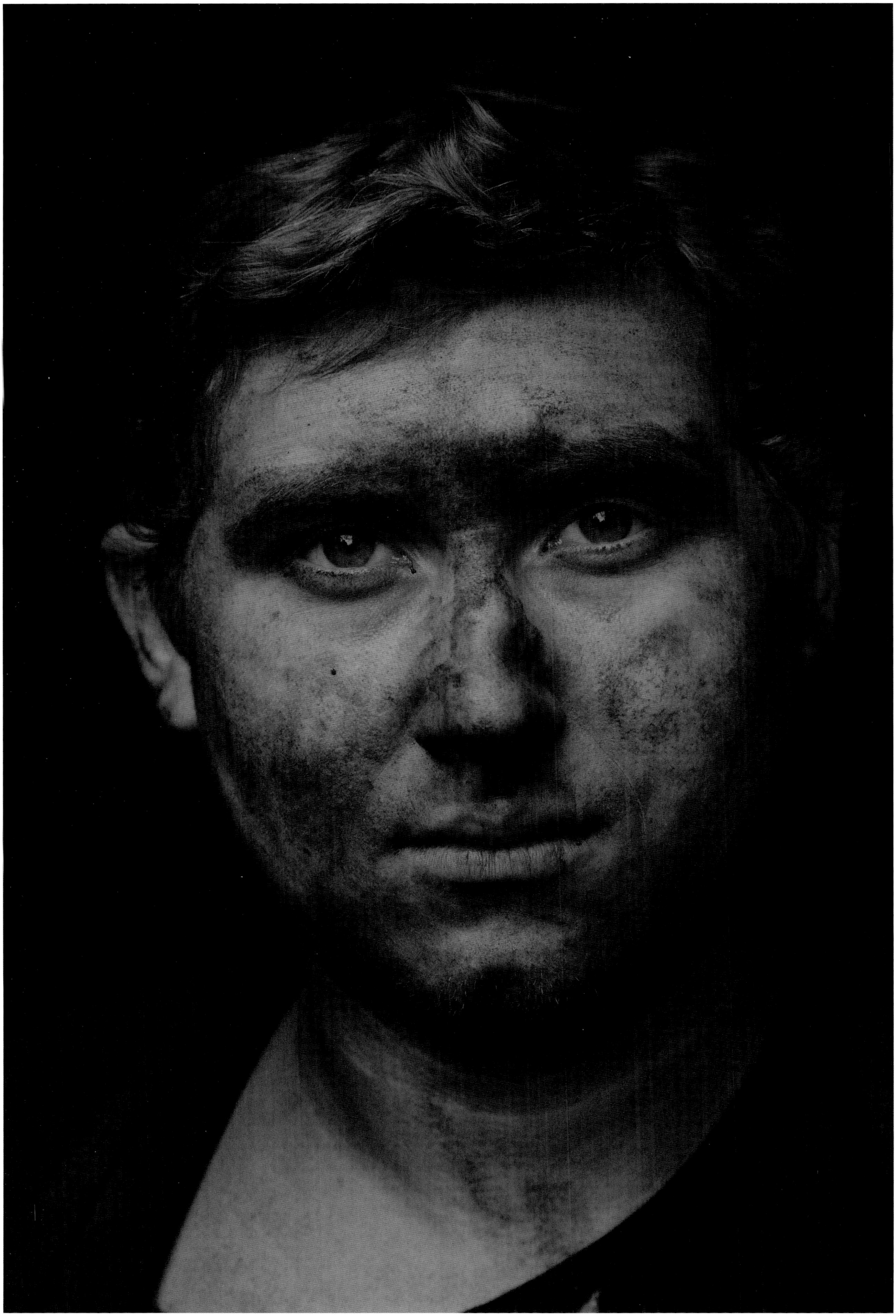

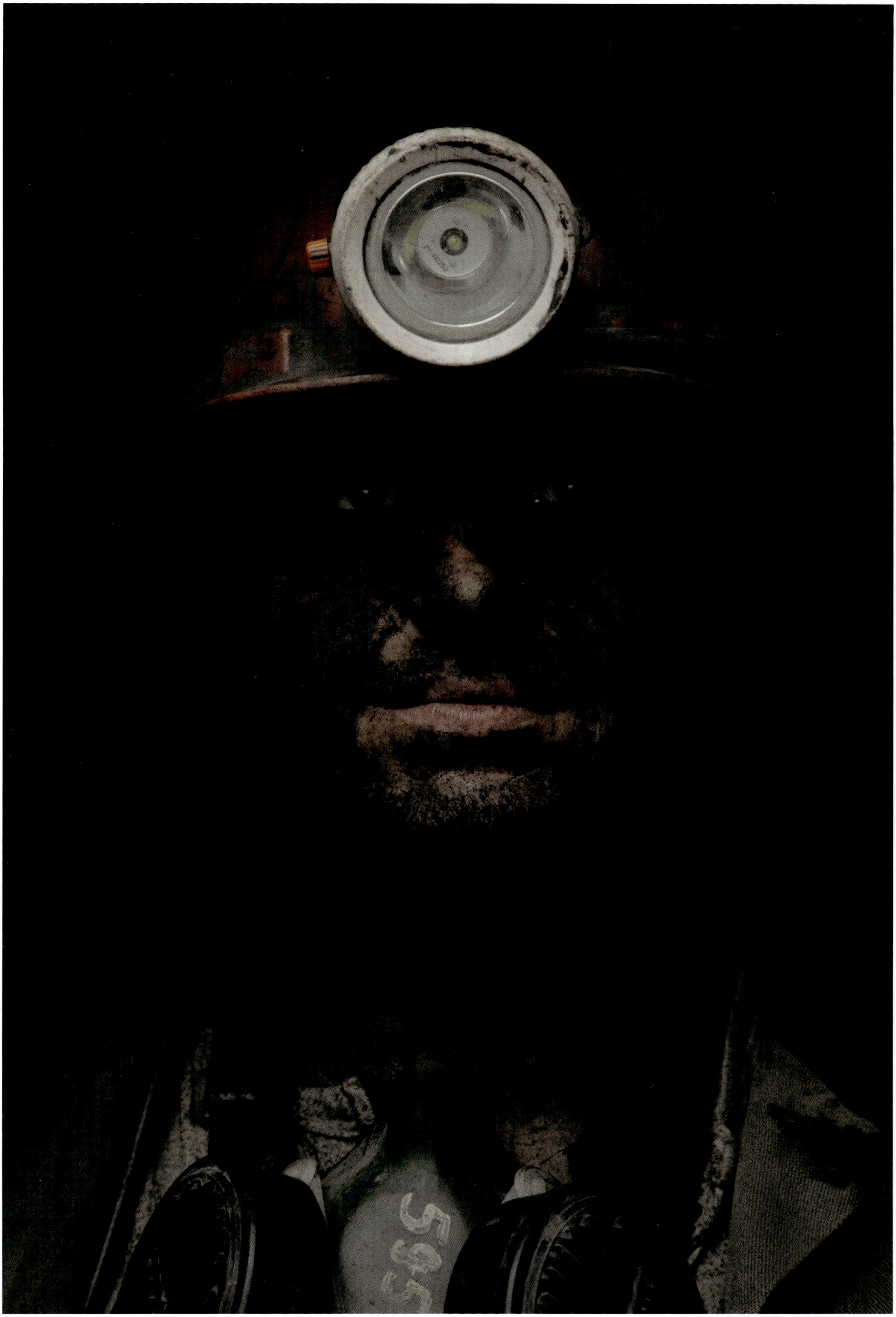

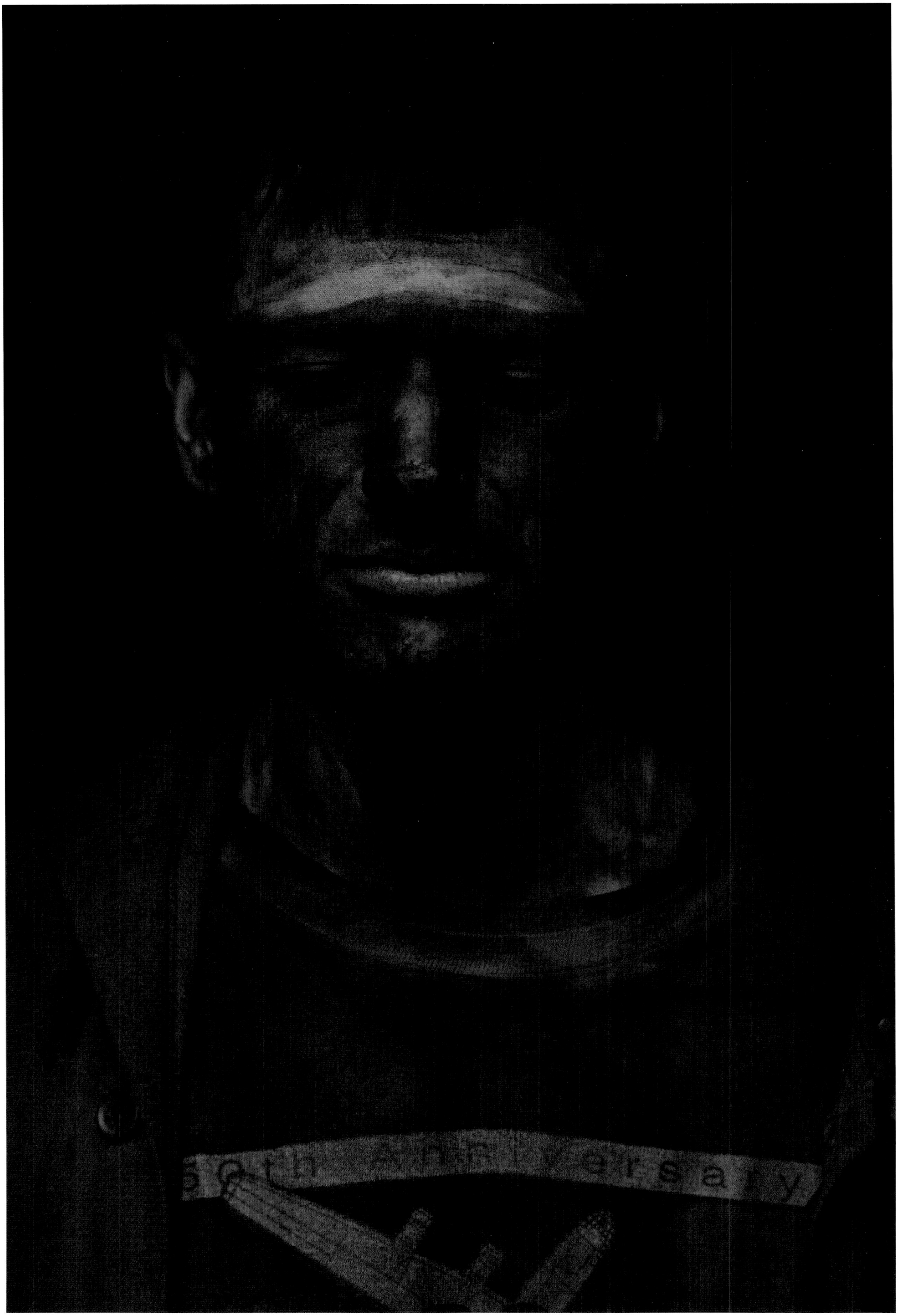
50th Anniversary

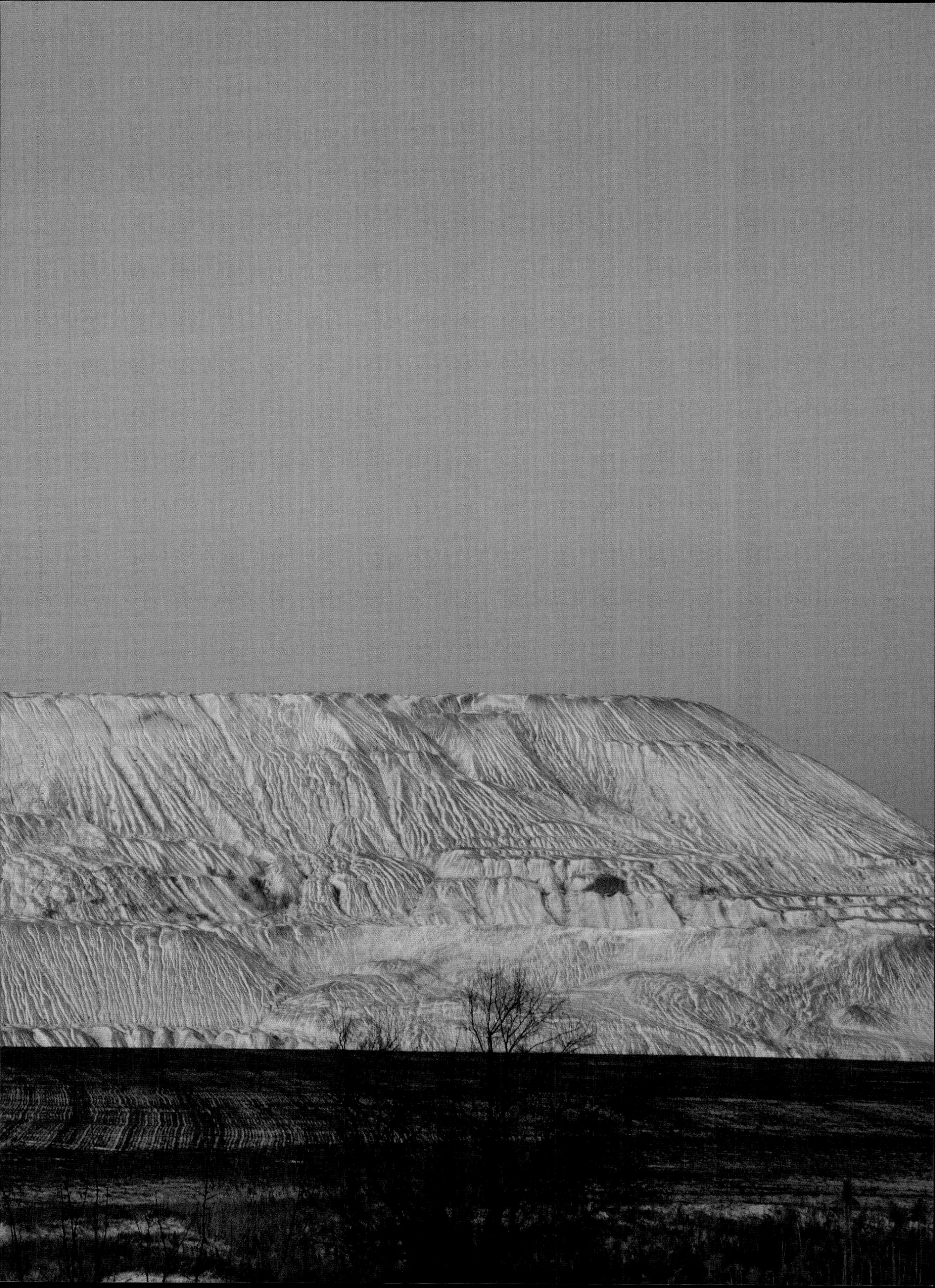

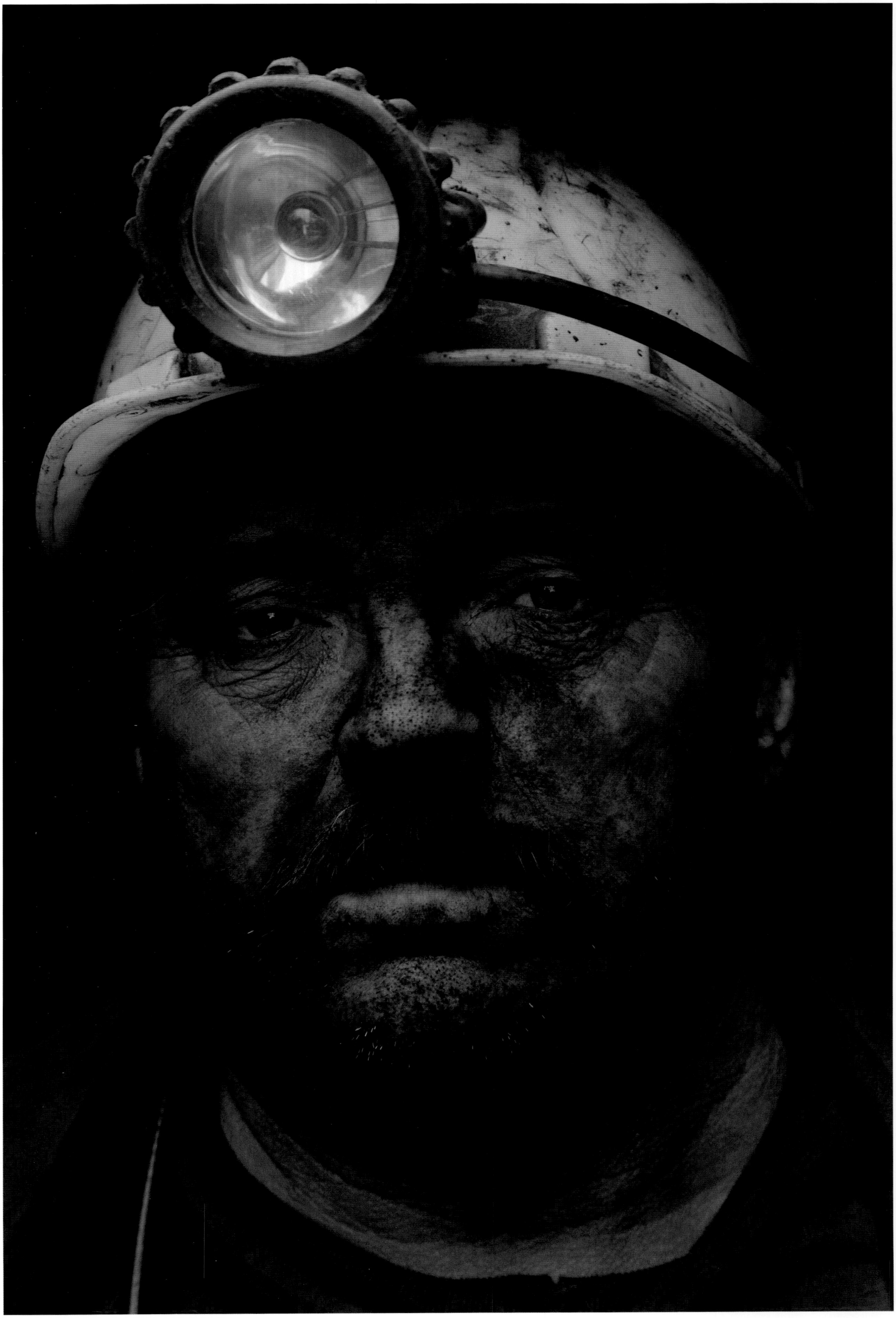

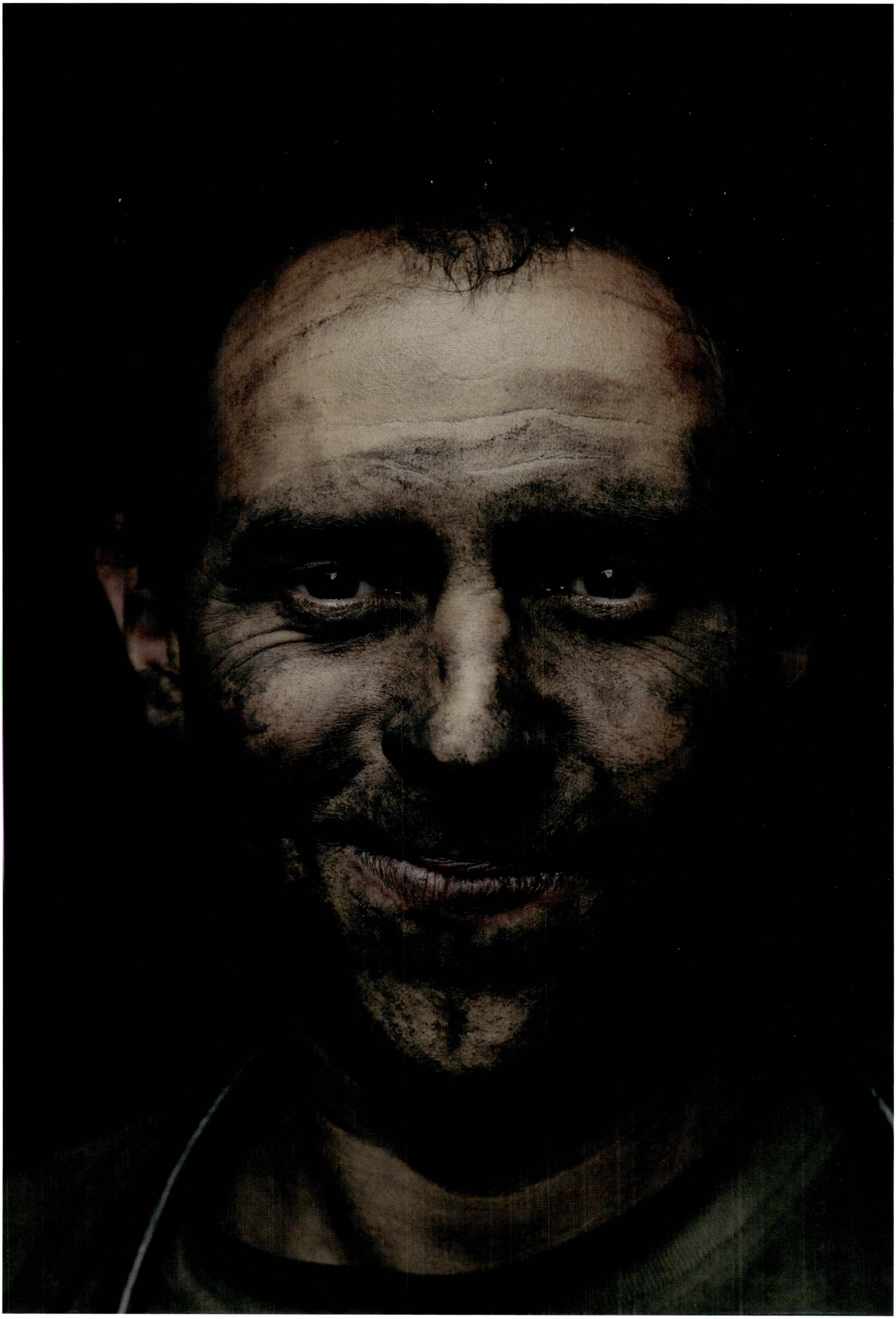

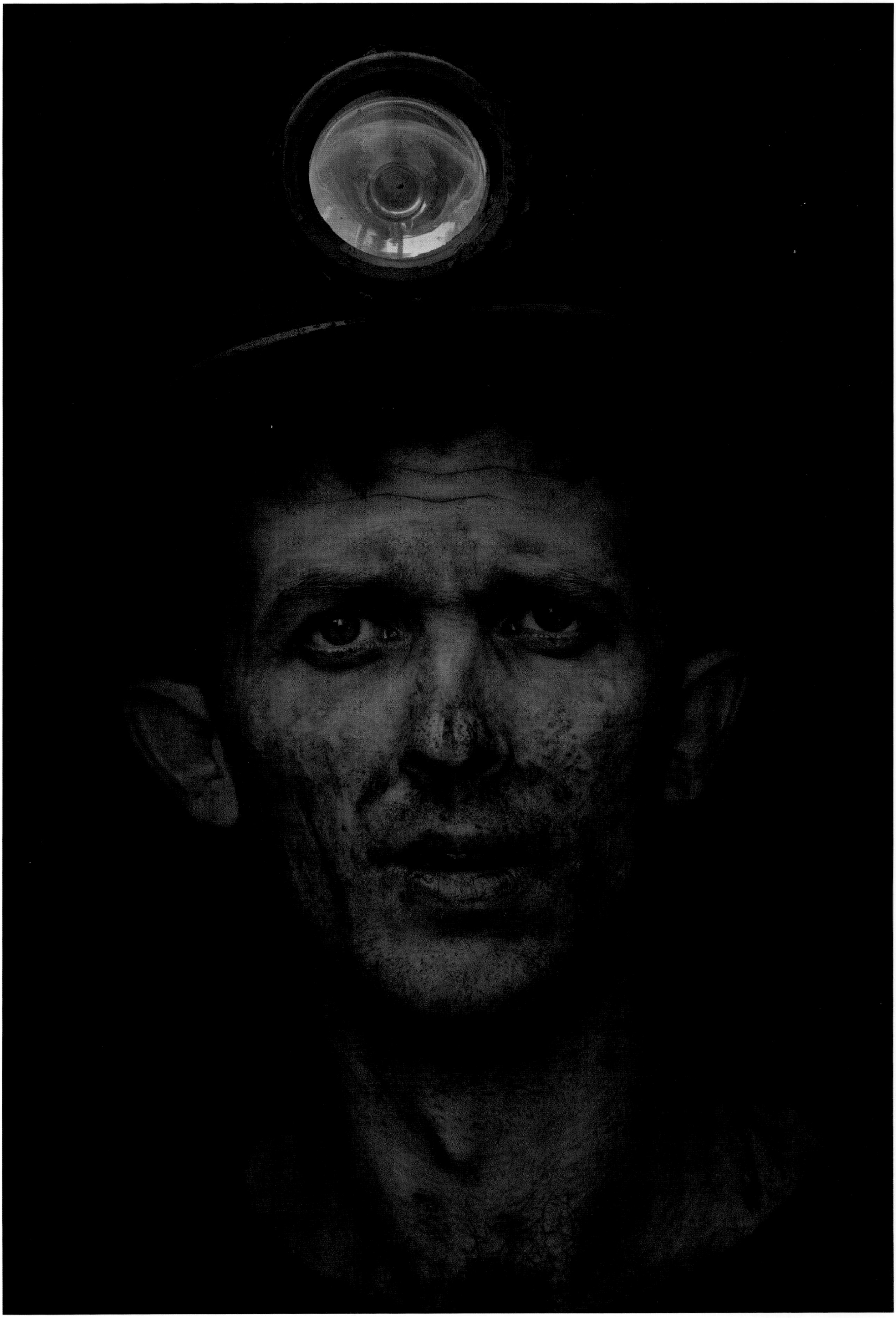

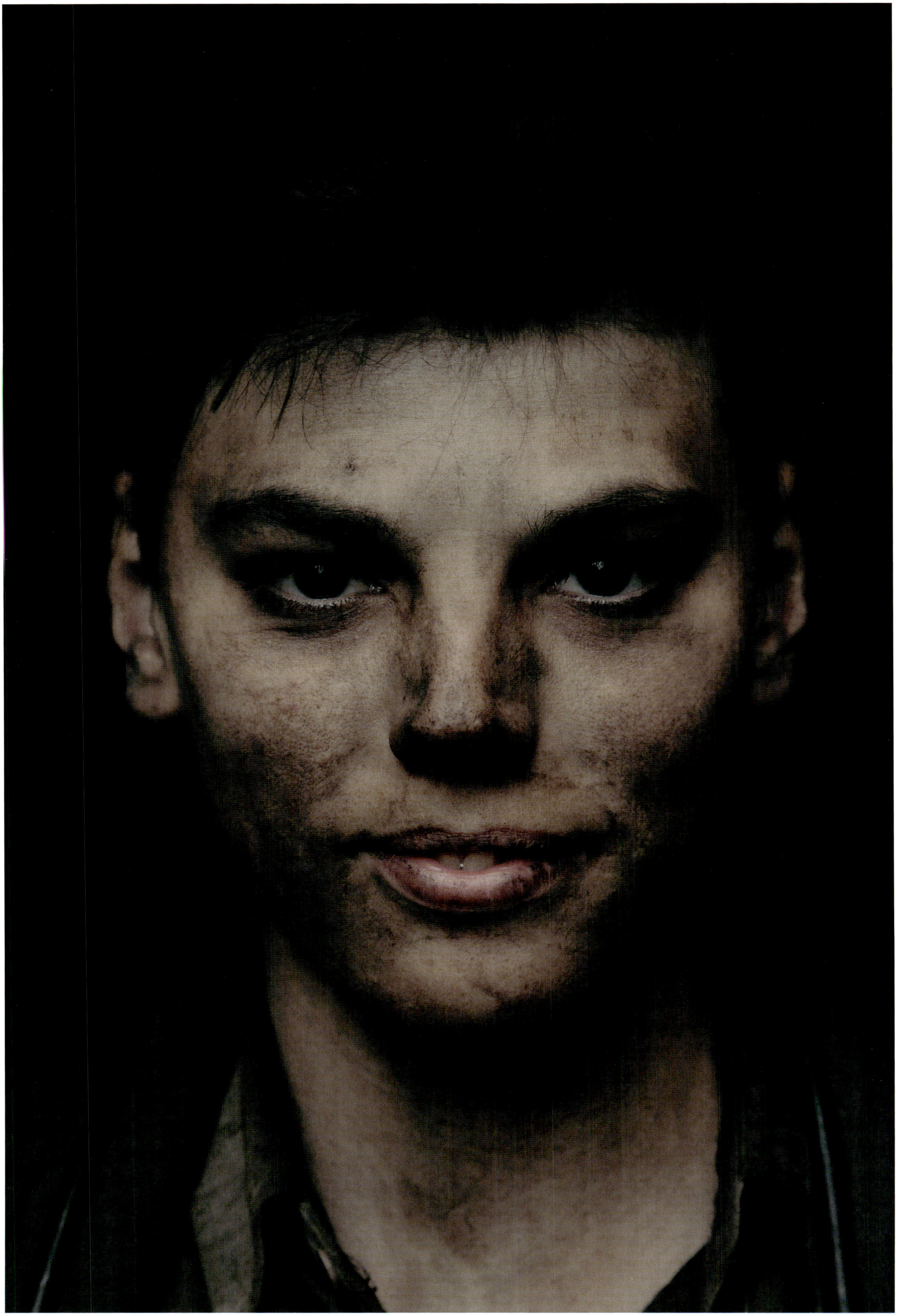

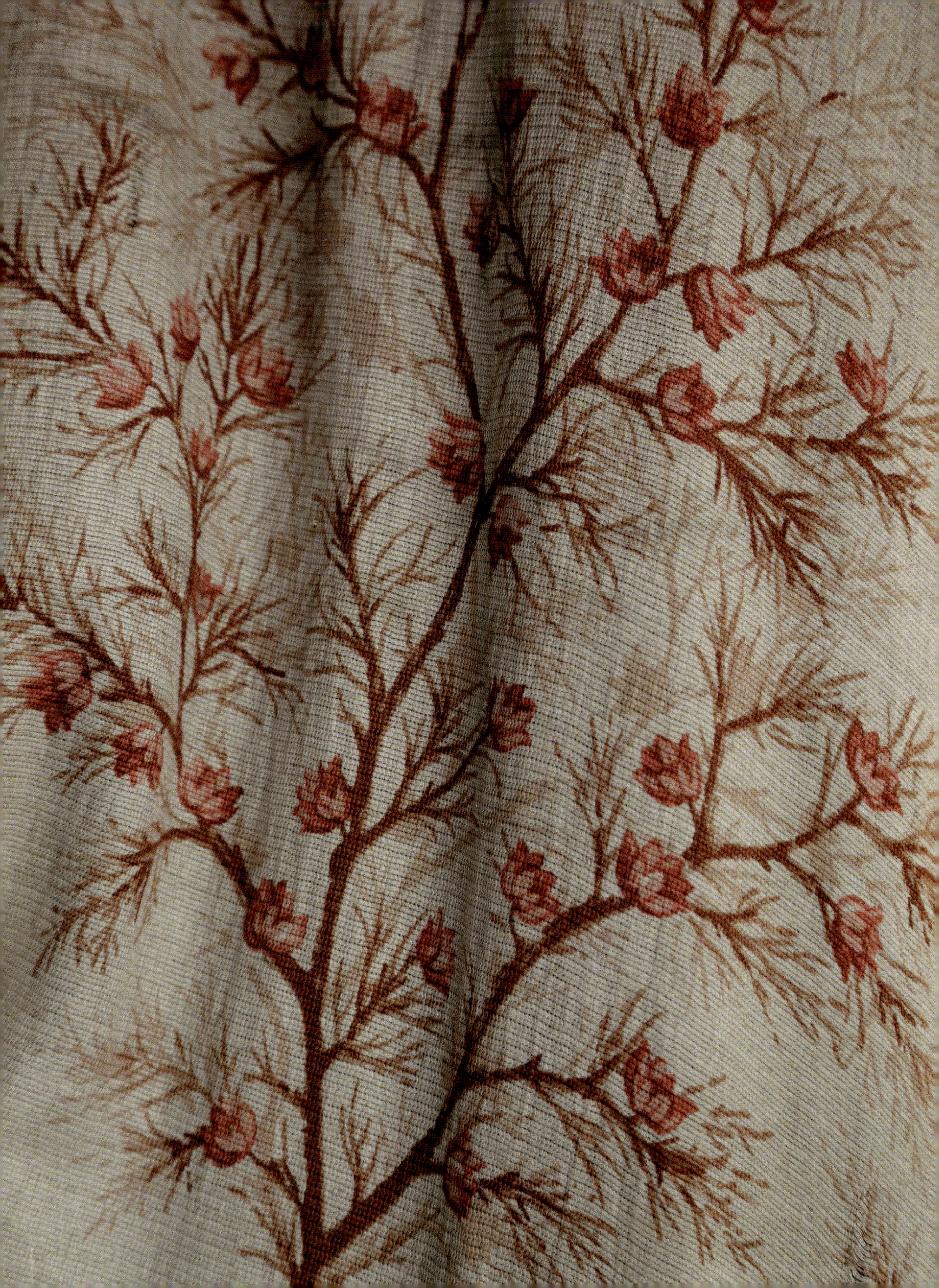

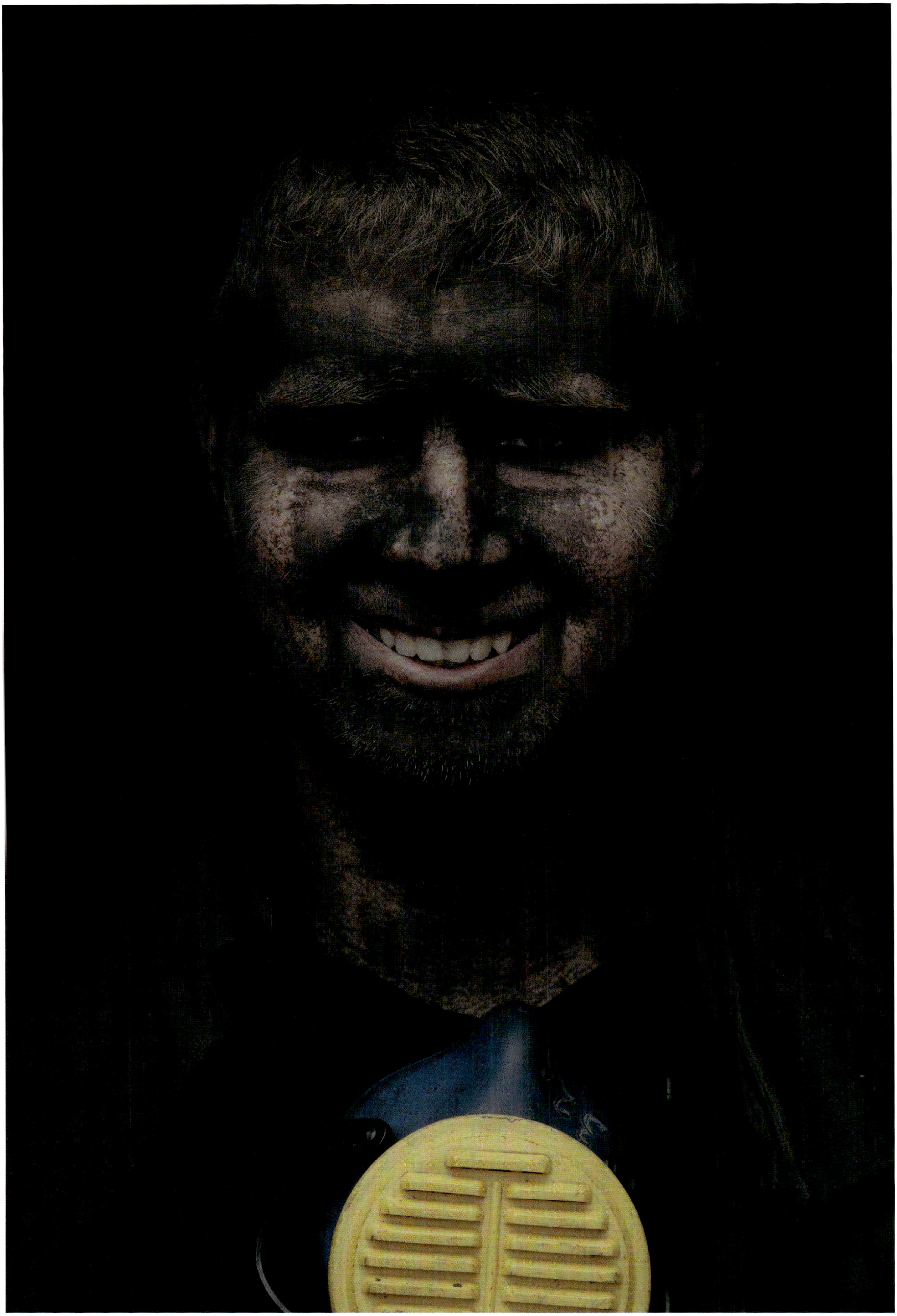

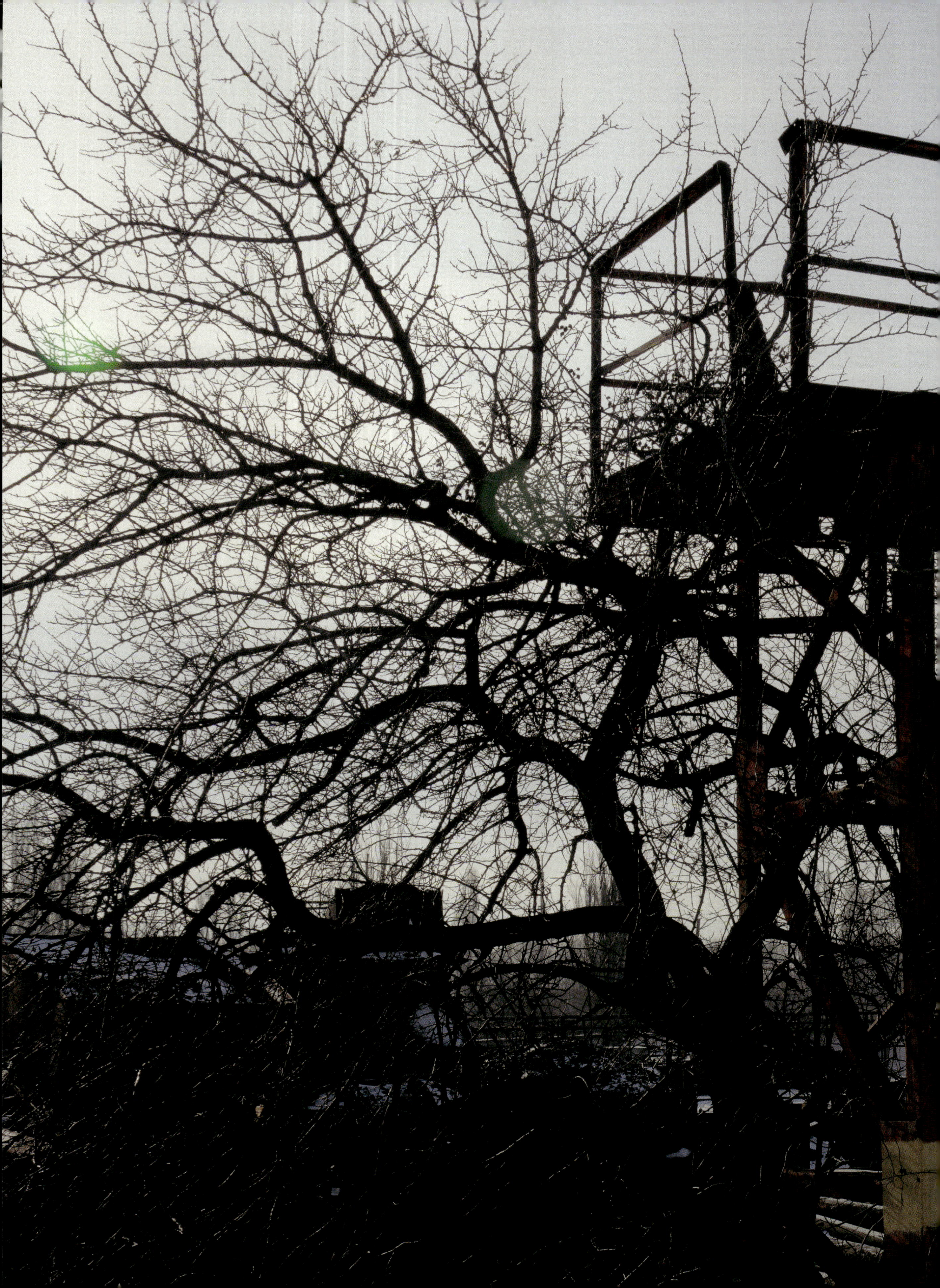

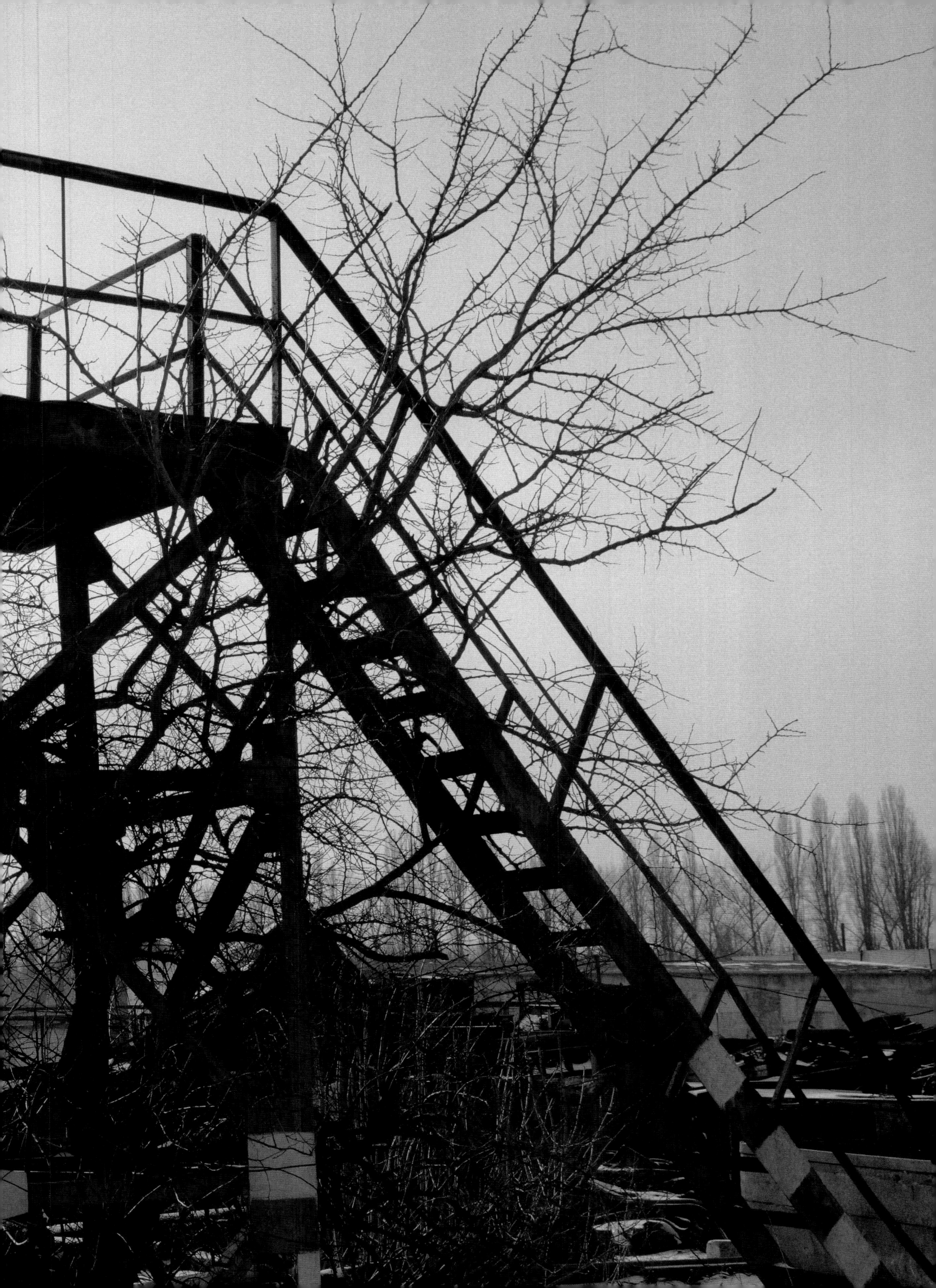

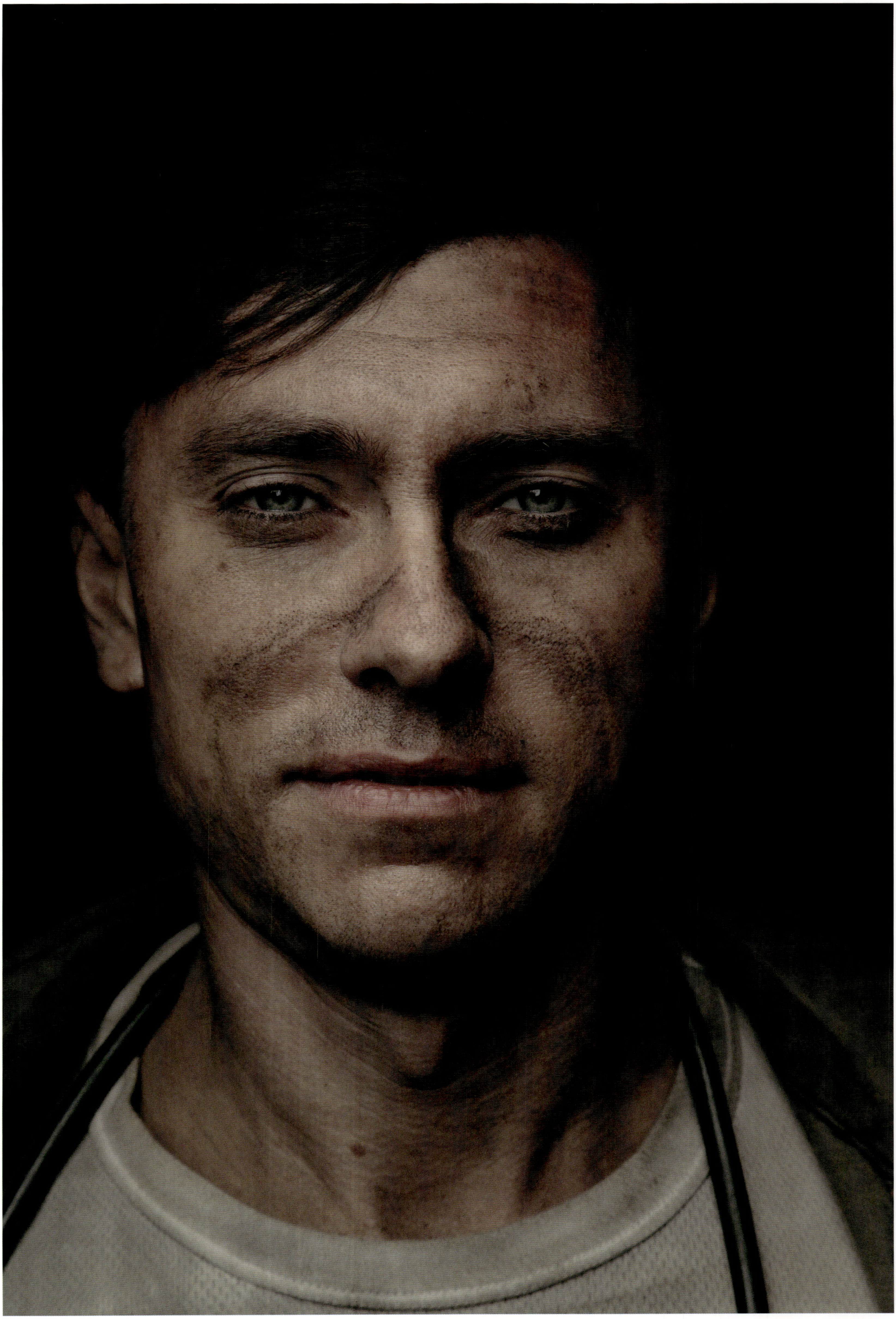

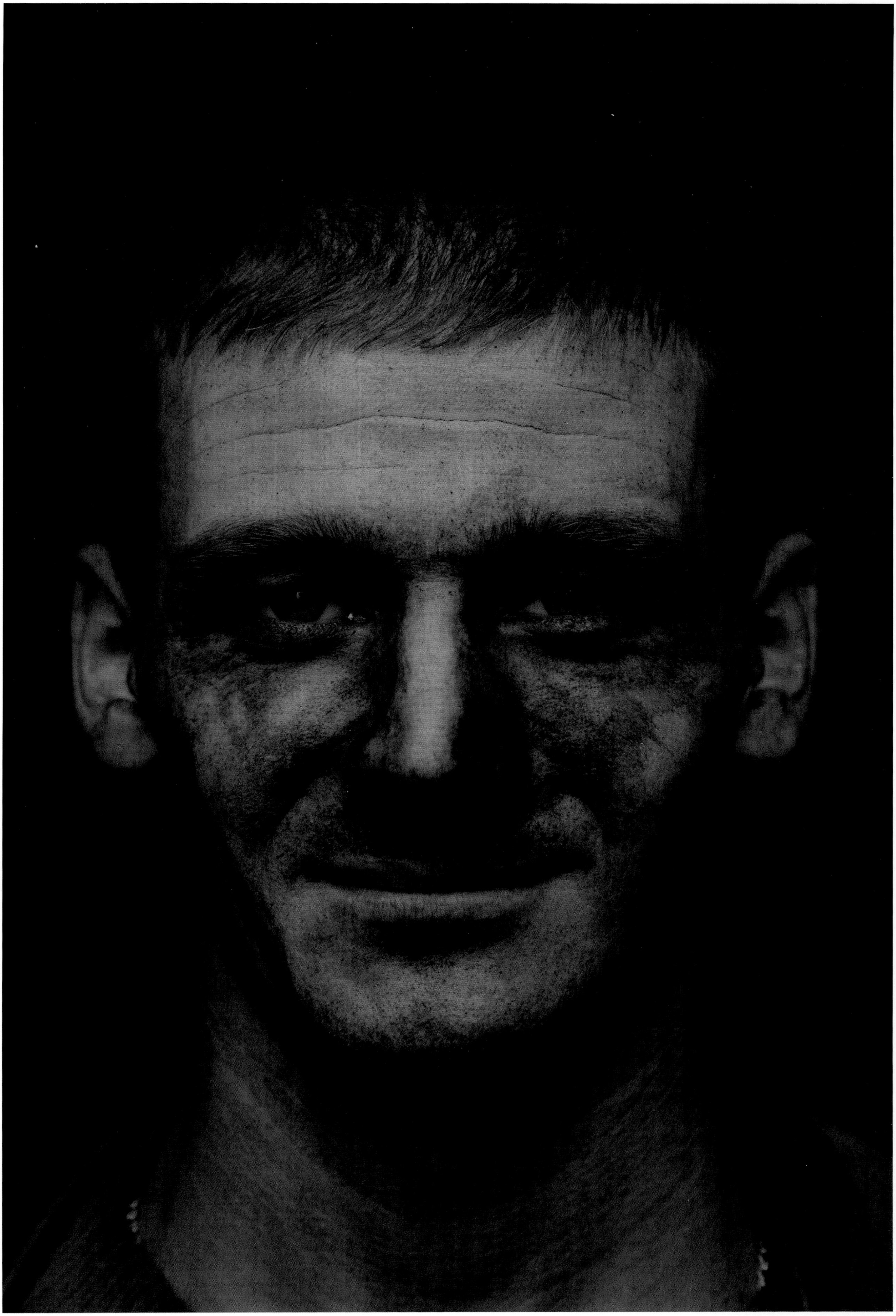

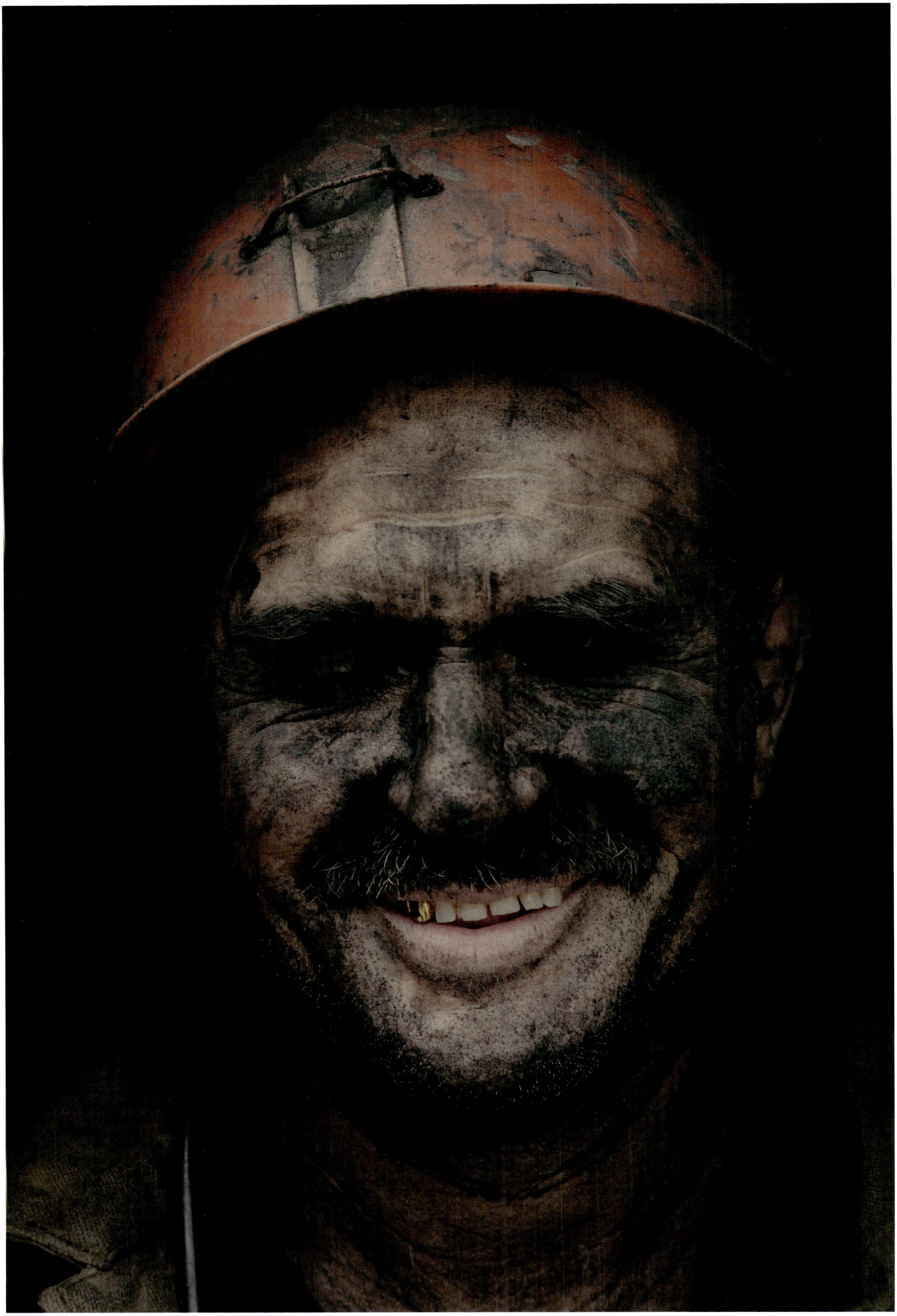

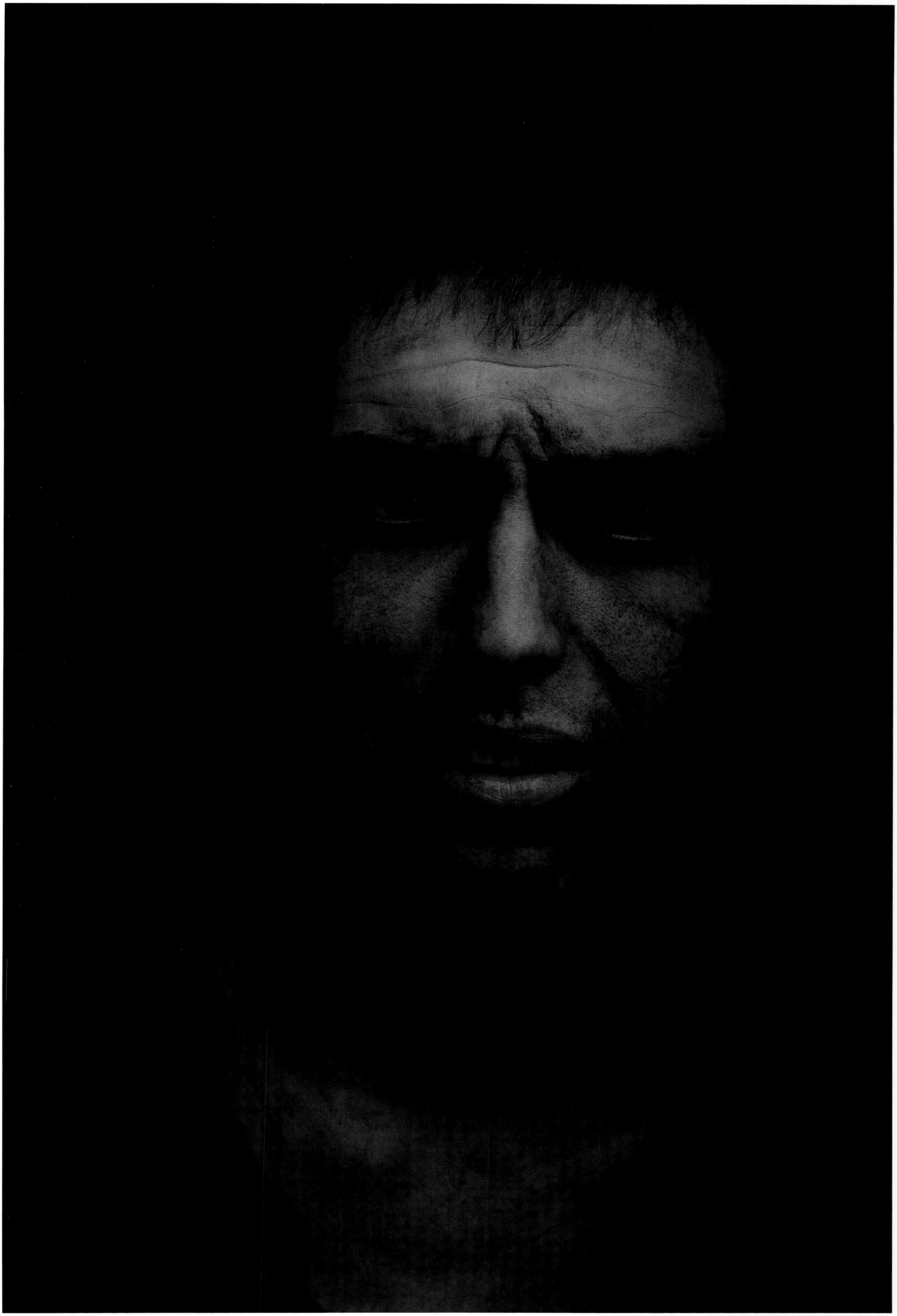

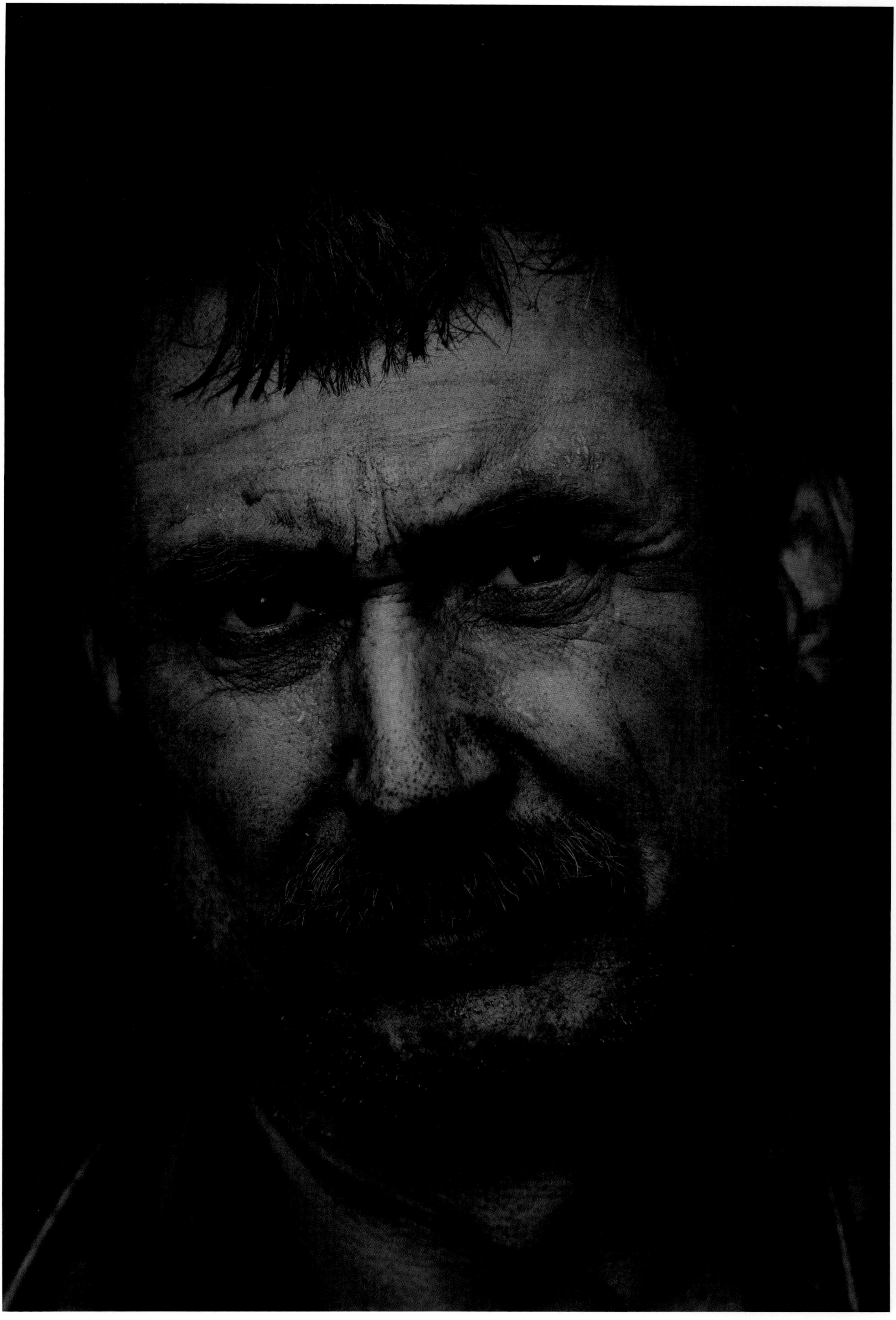

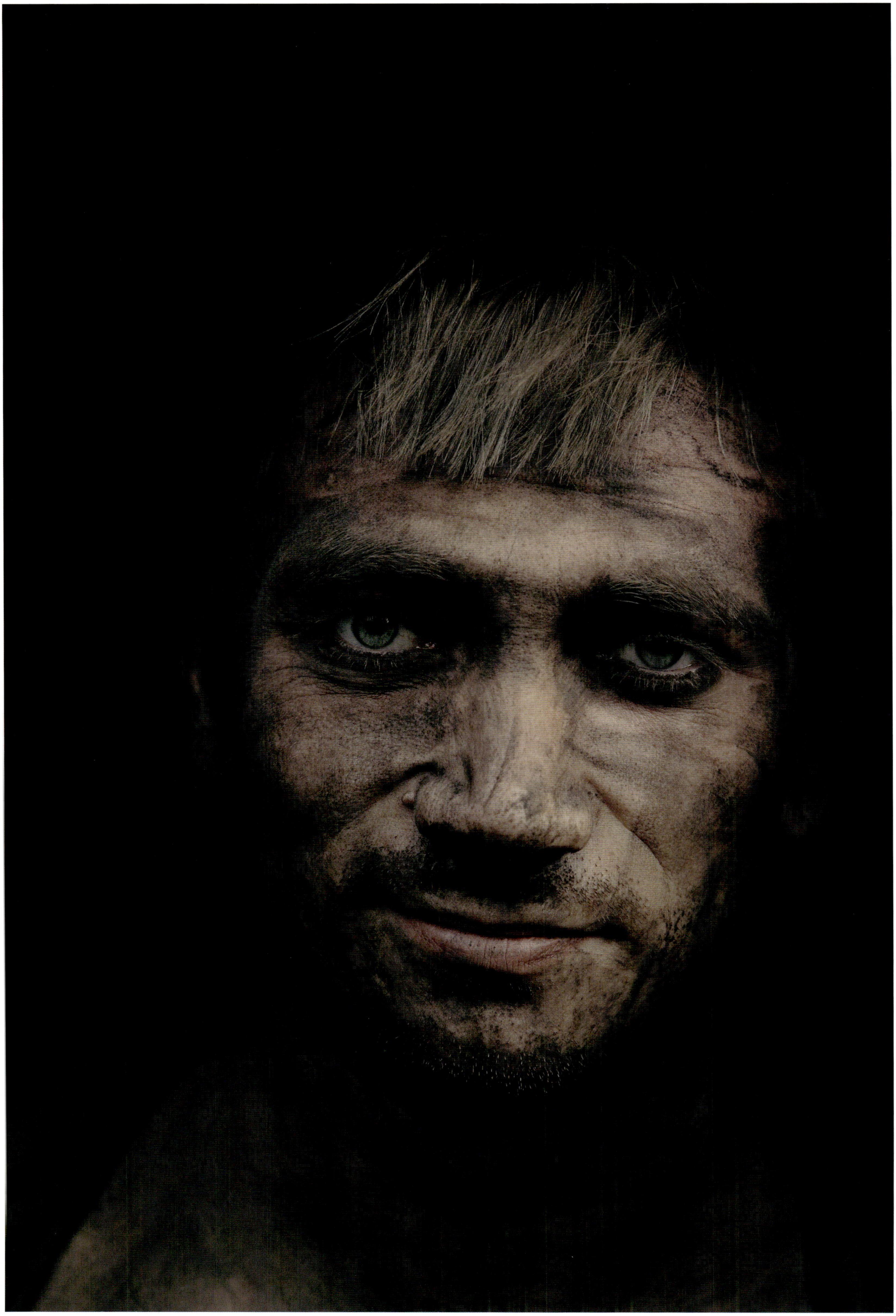

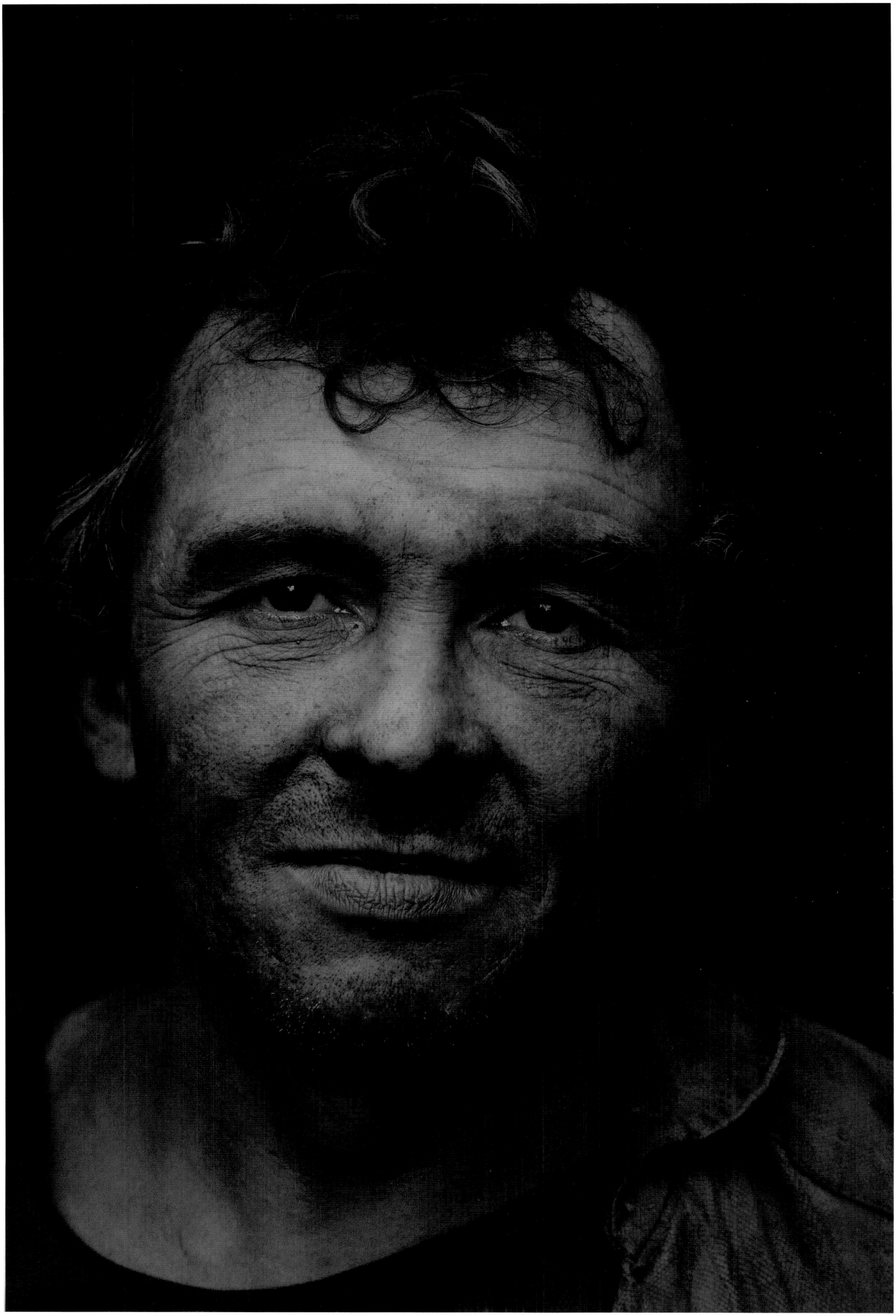

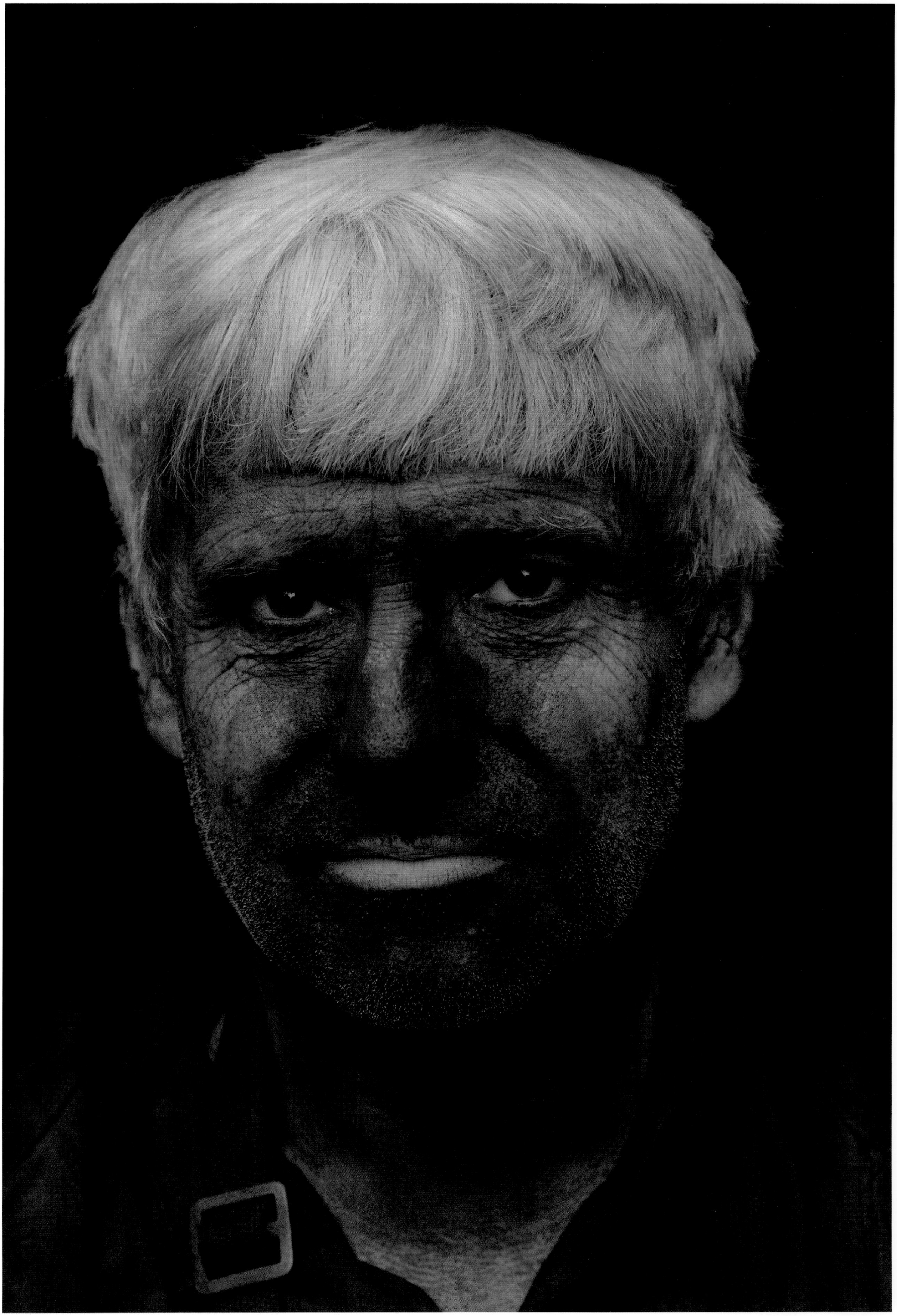

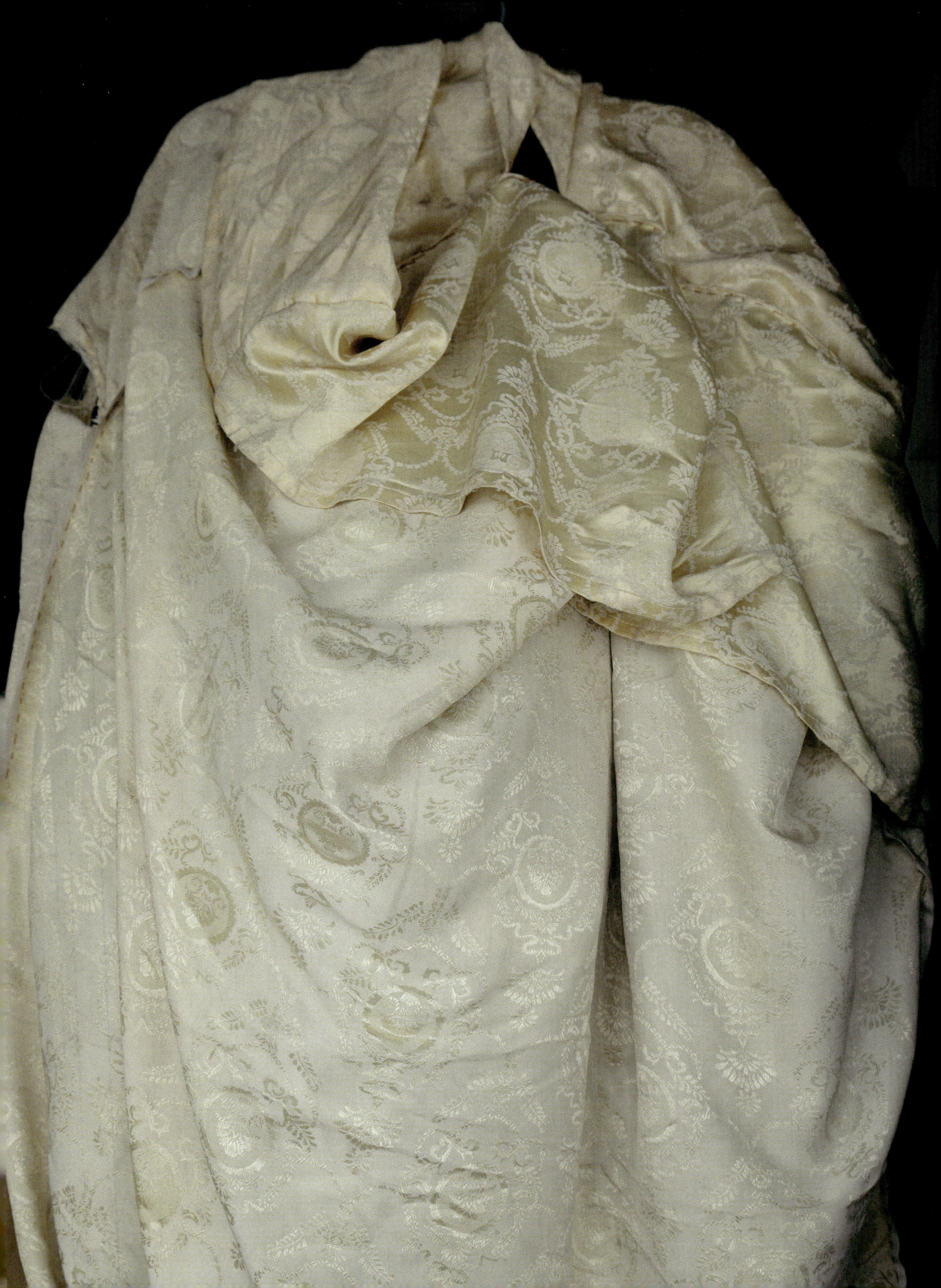

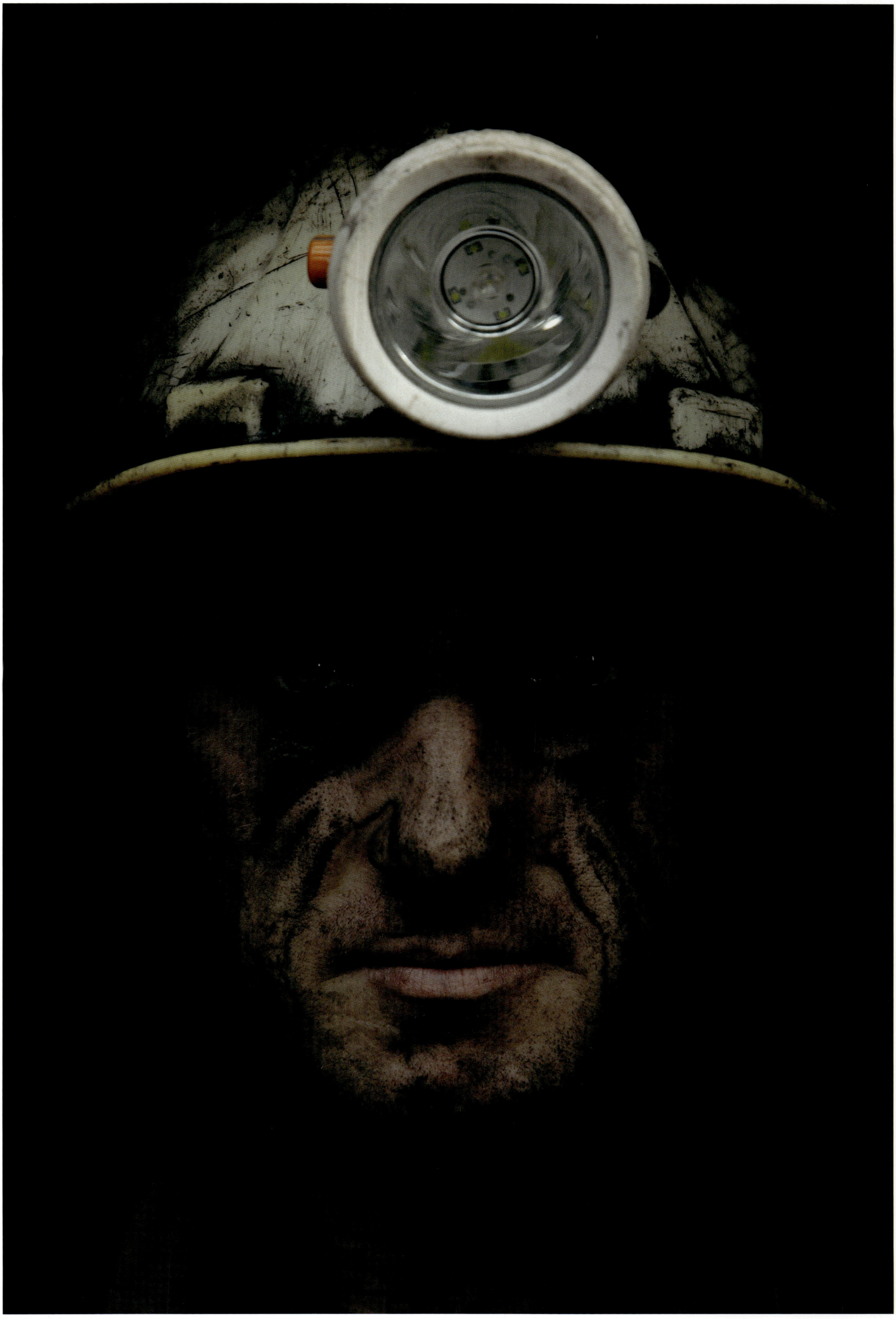

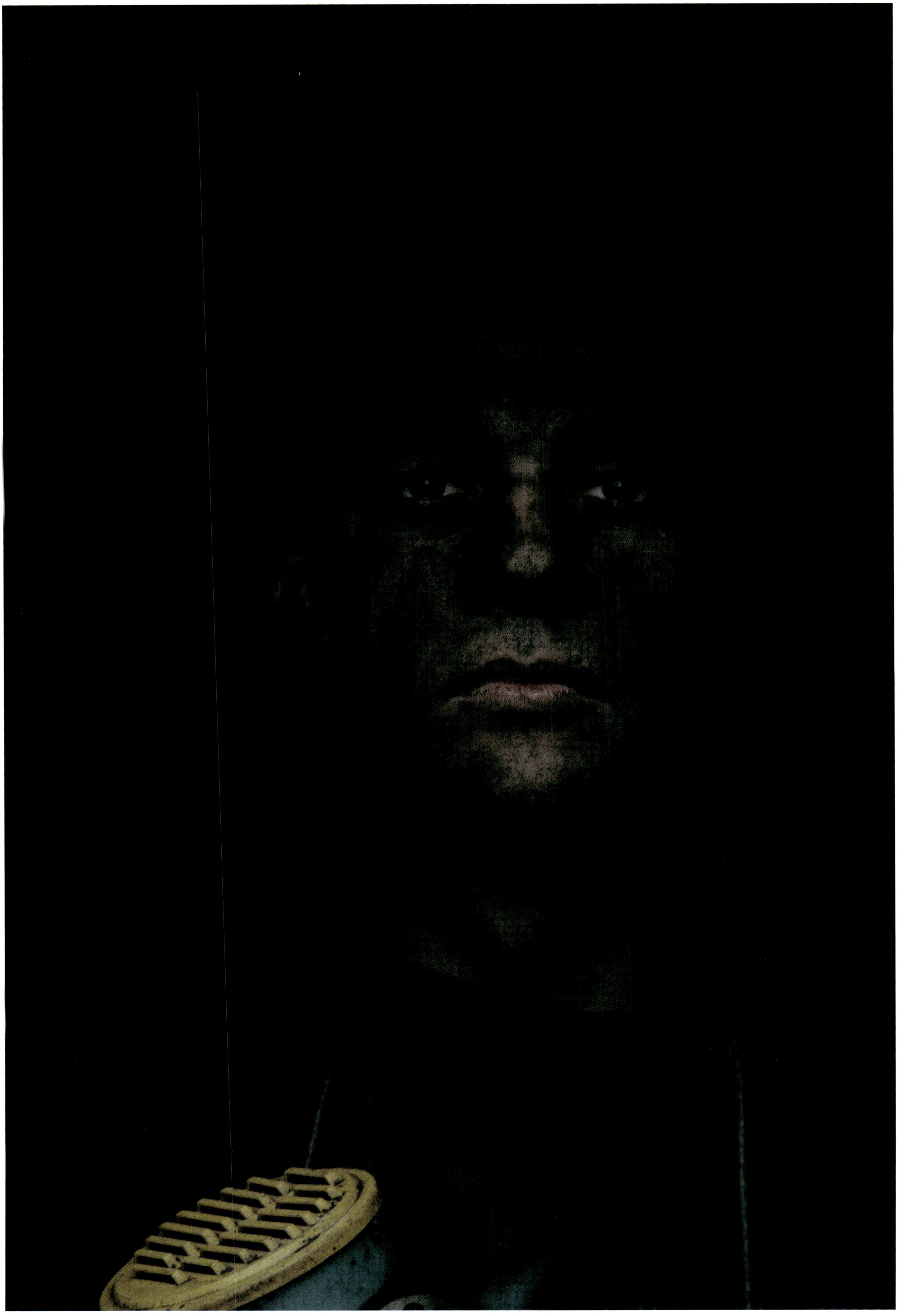

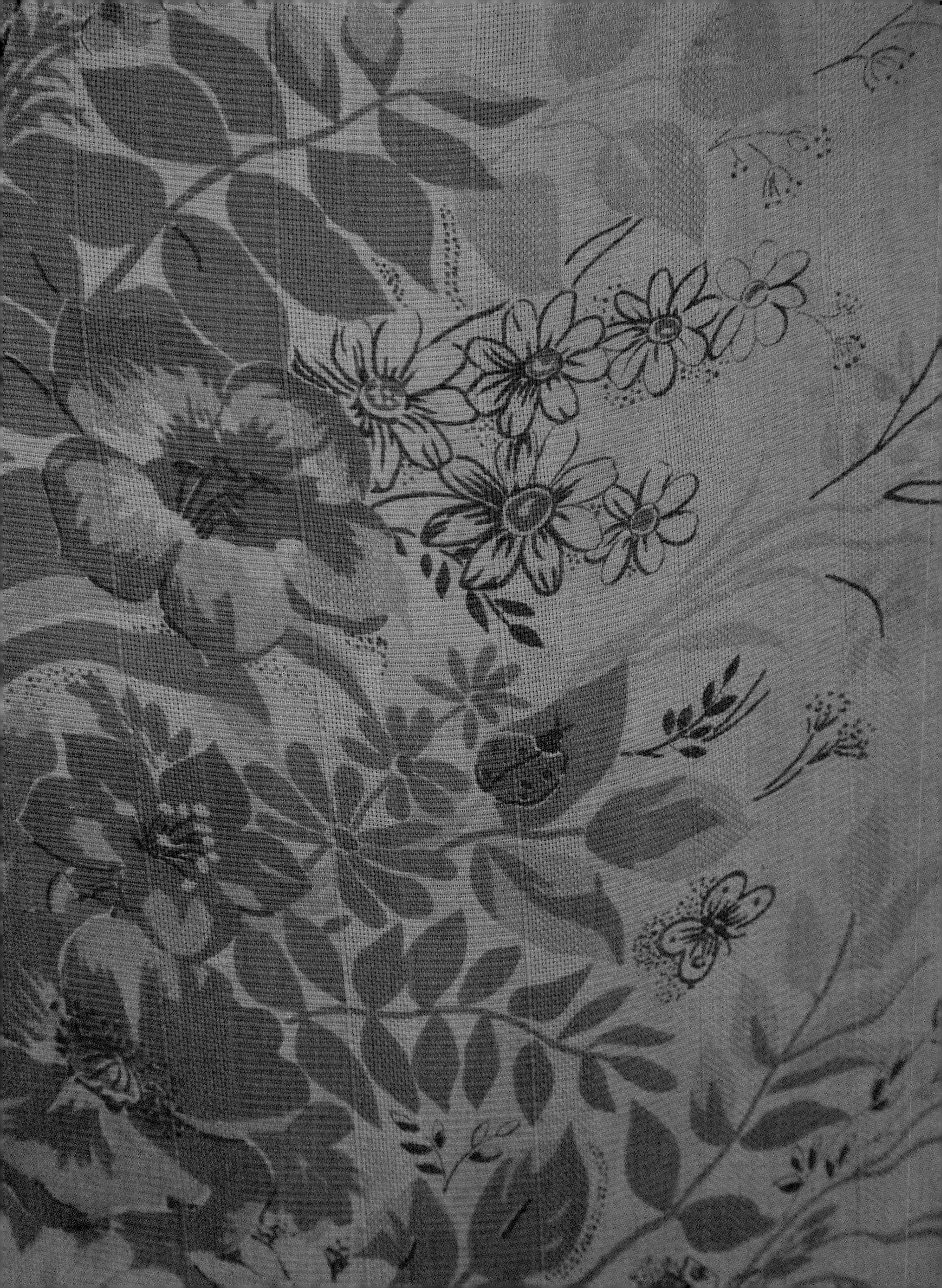

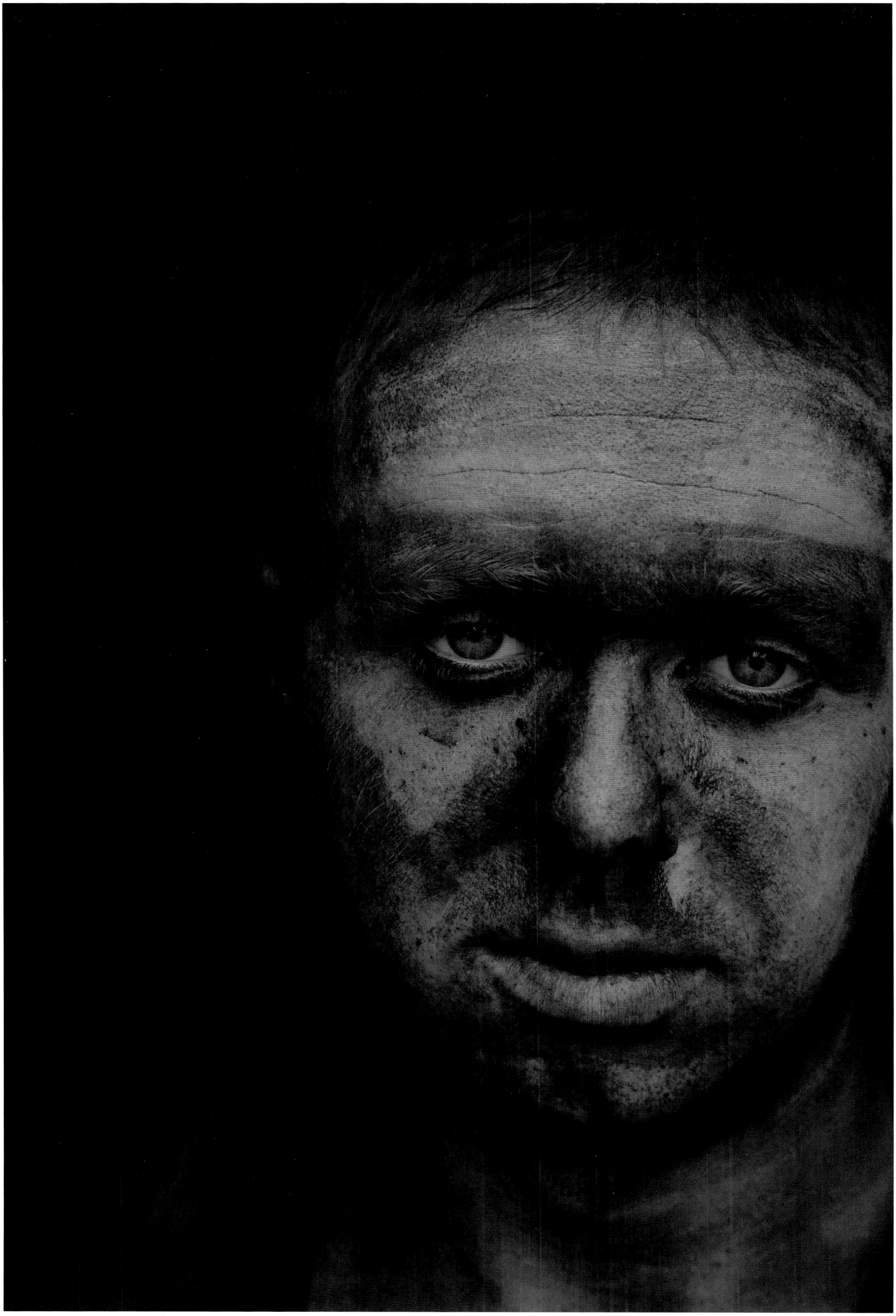

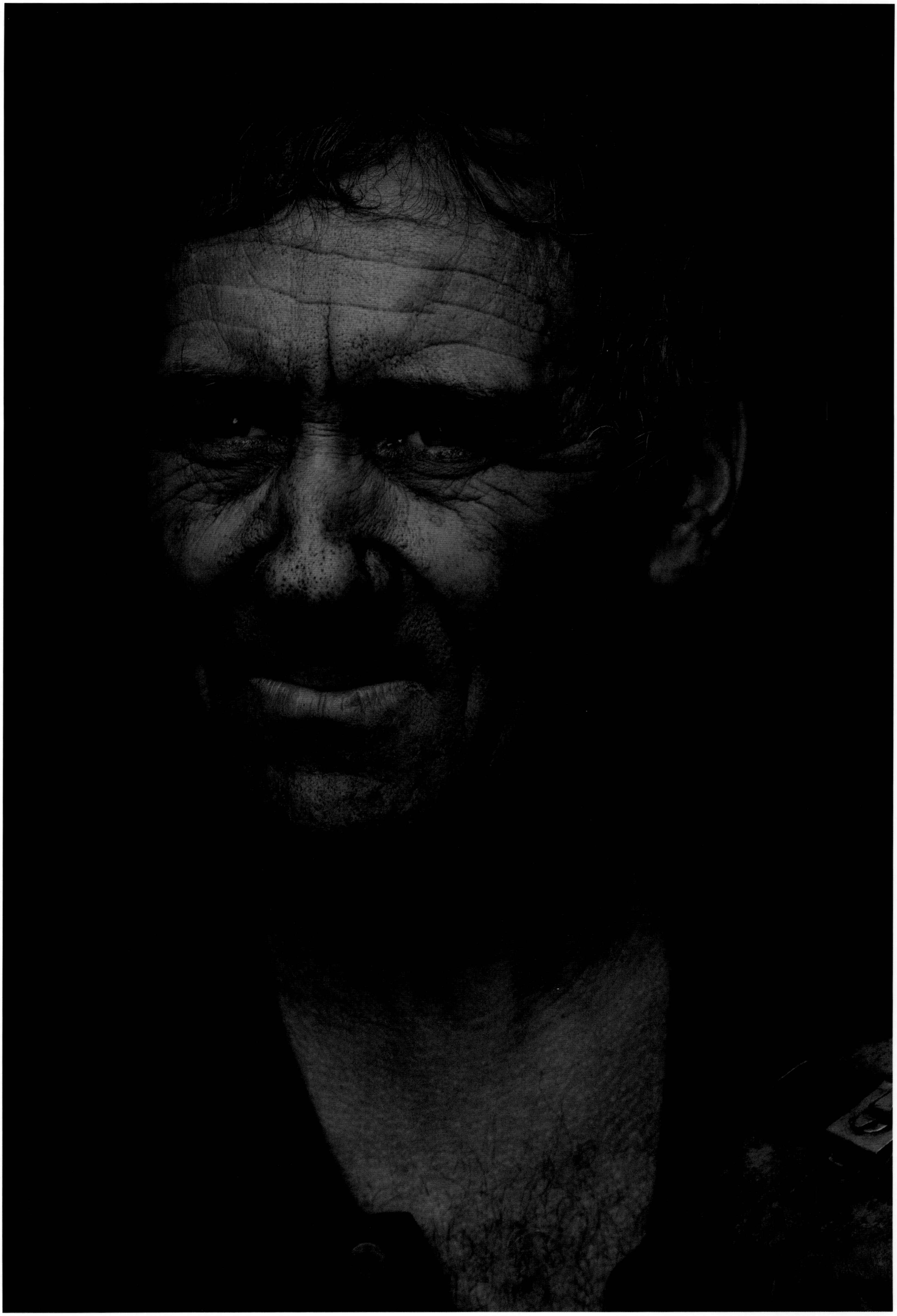

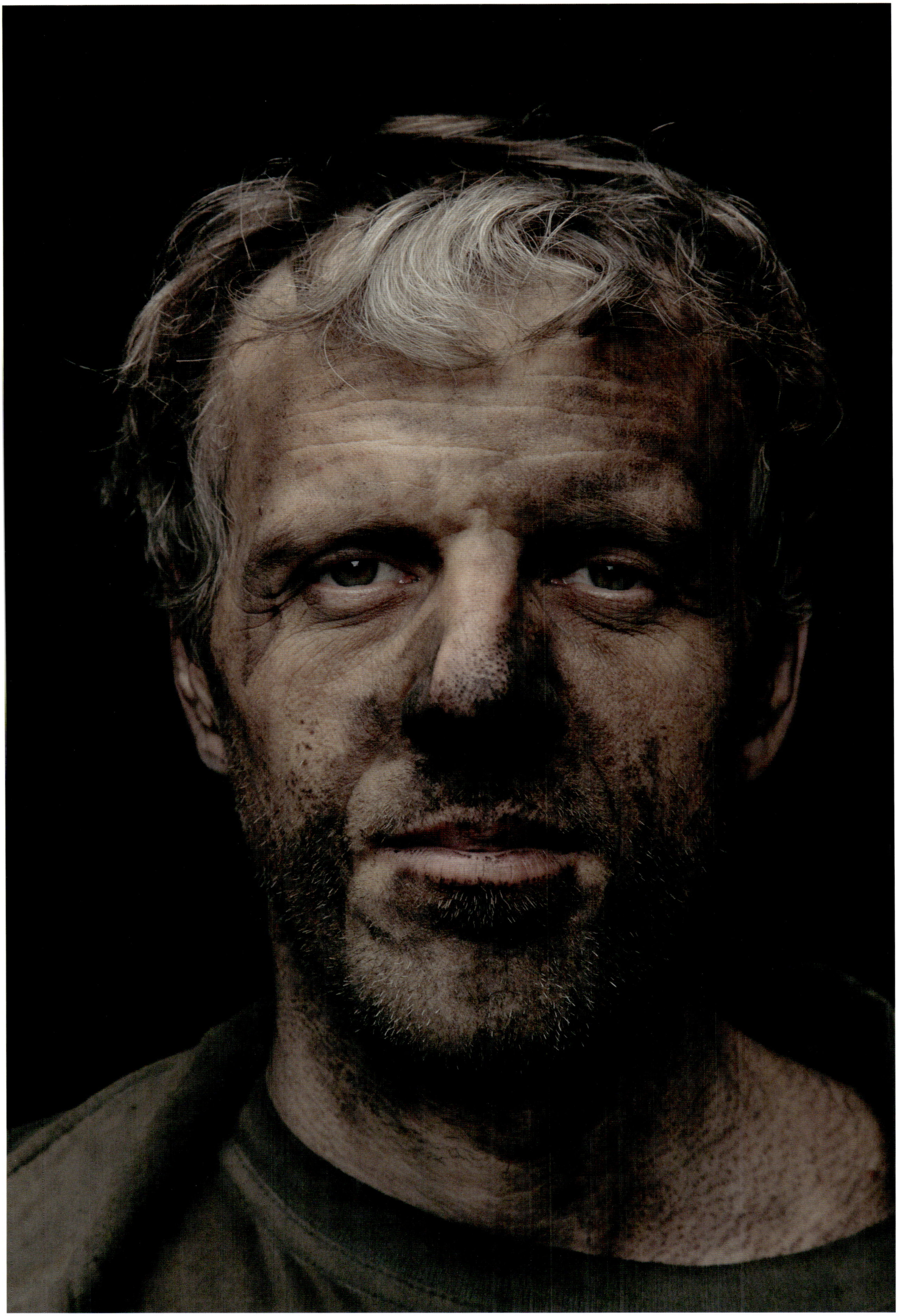

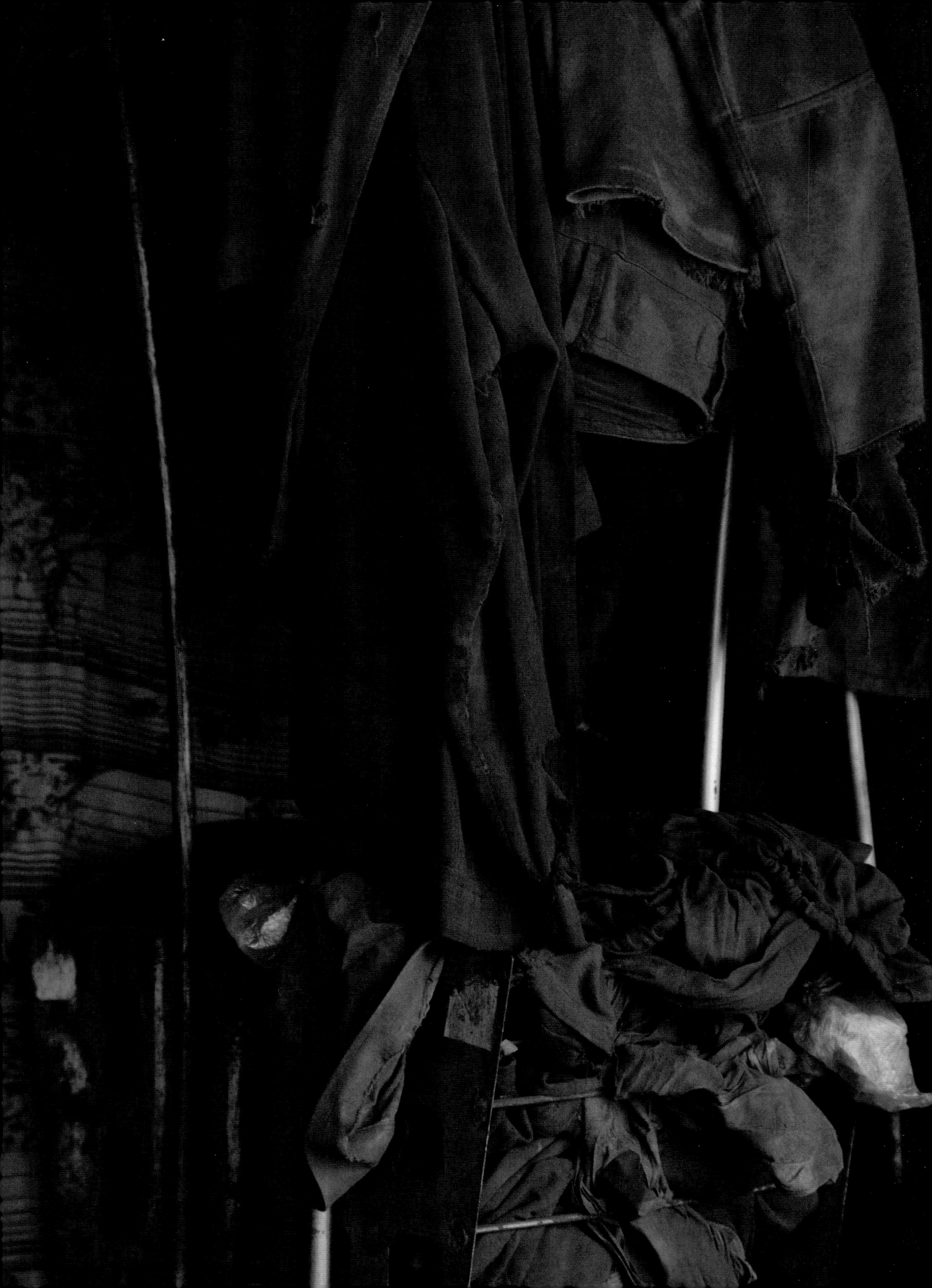

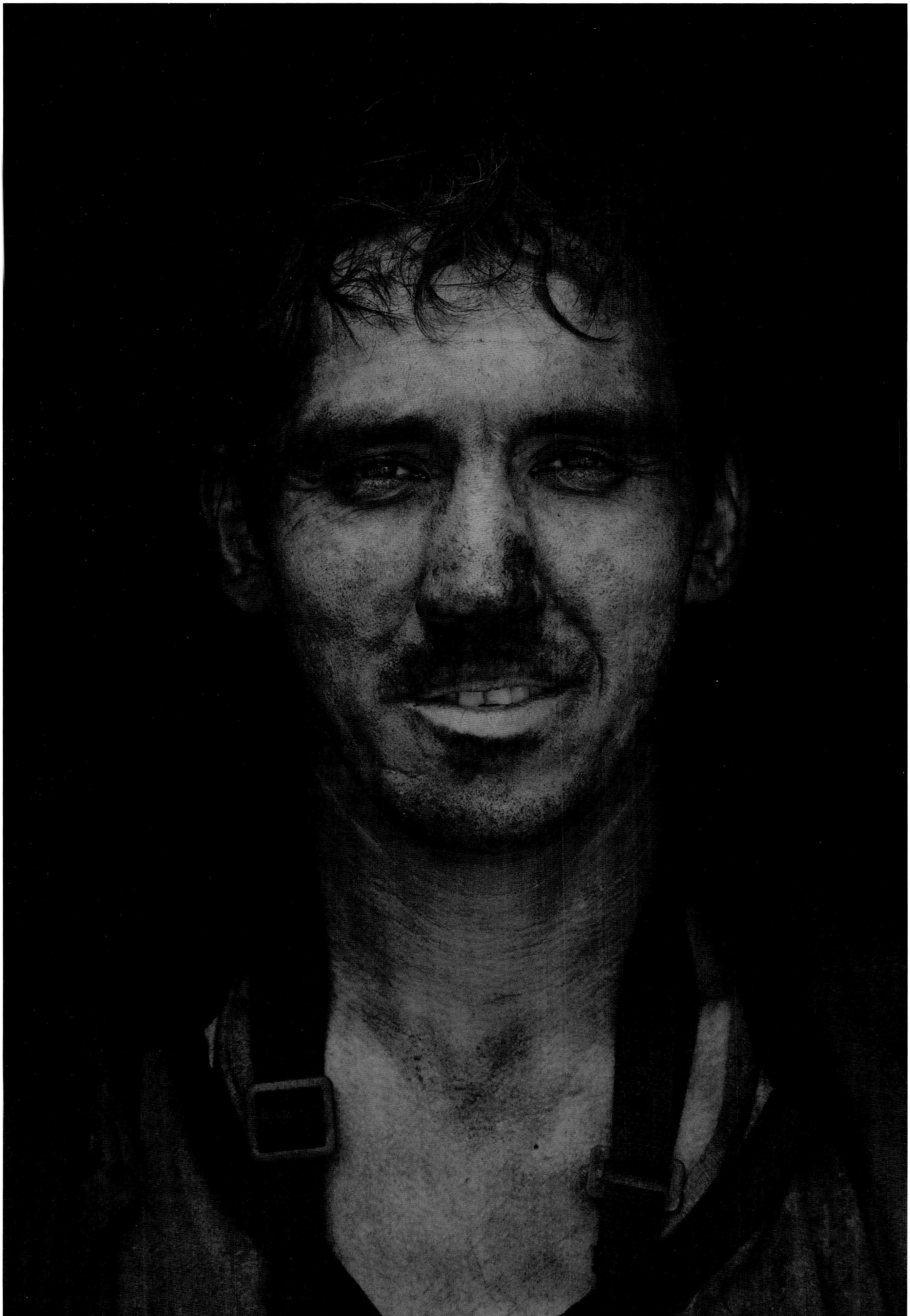

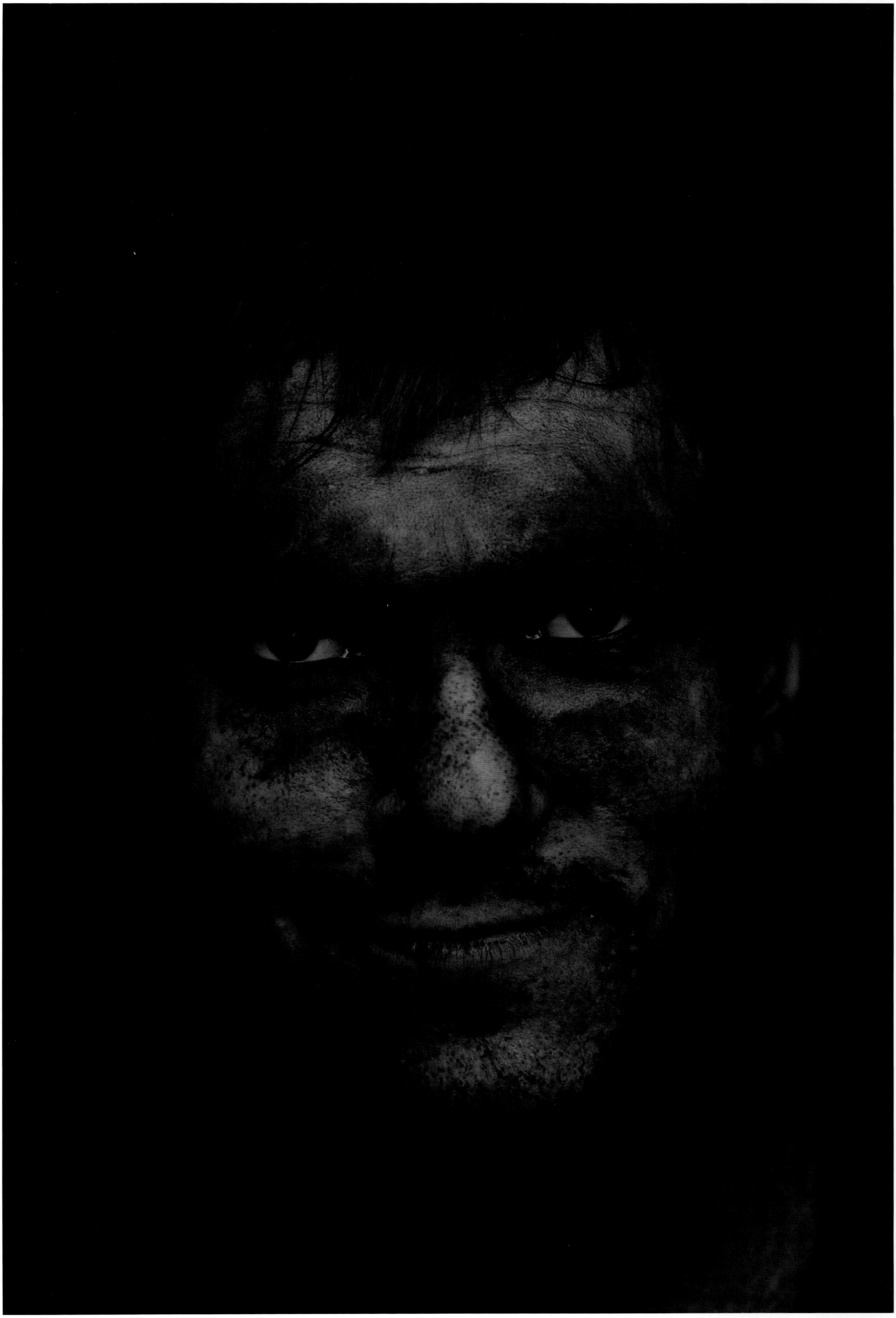

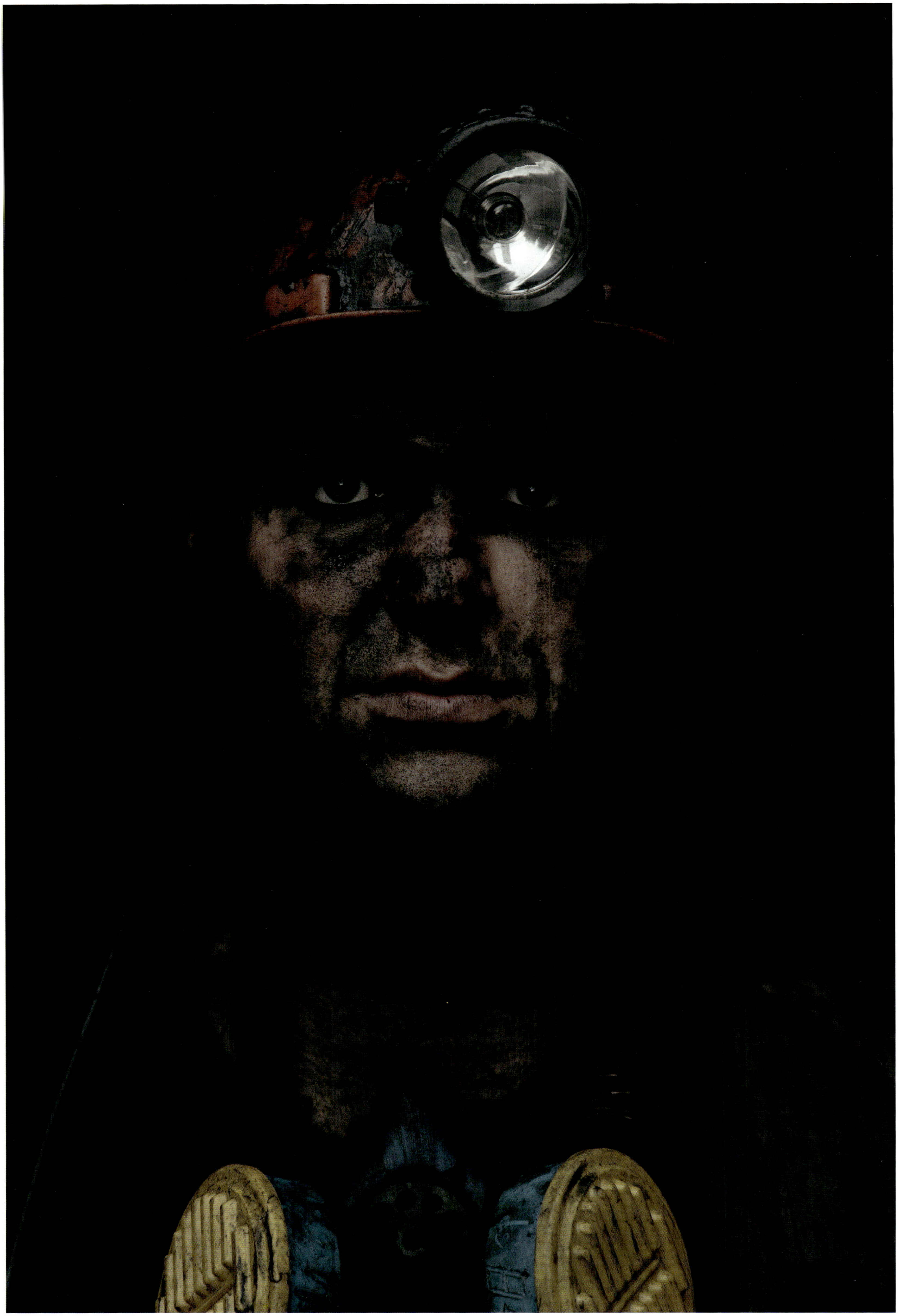

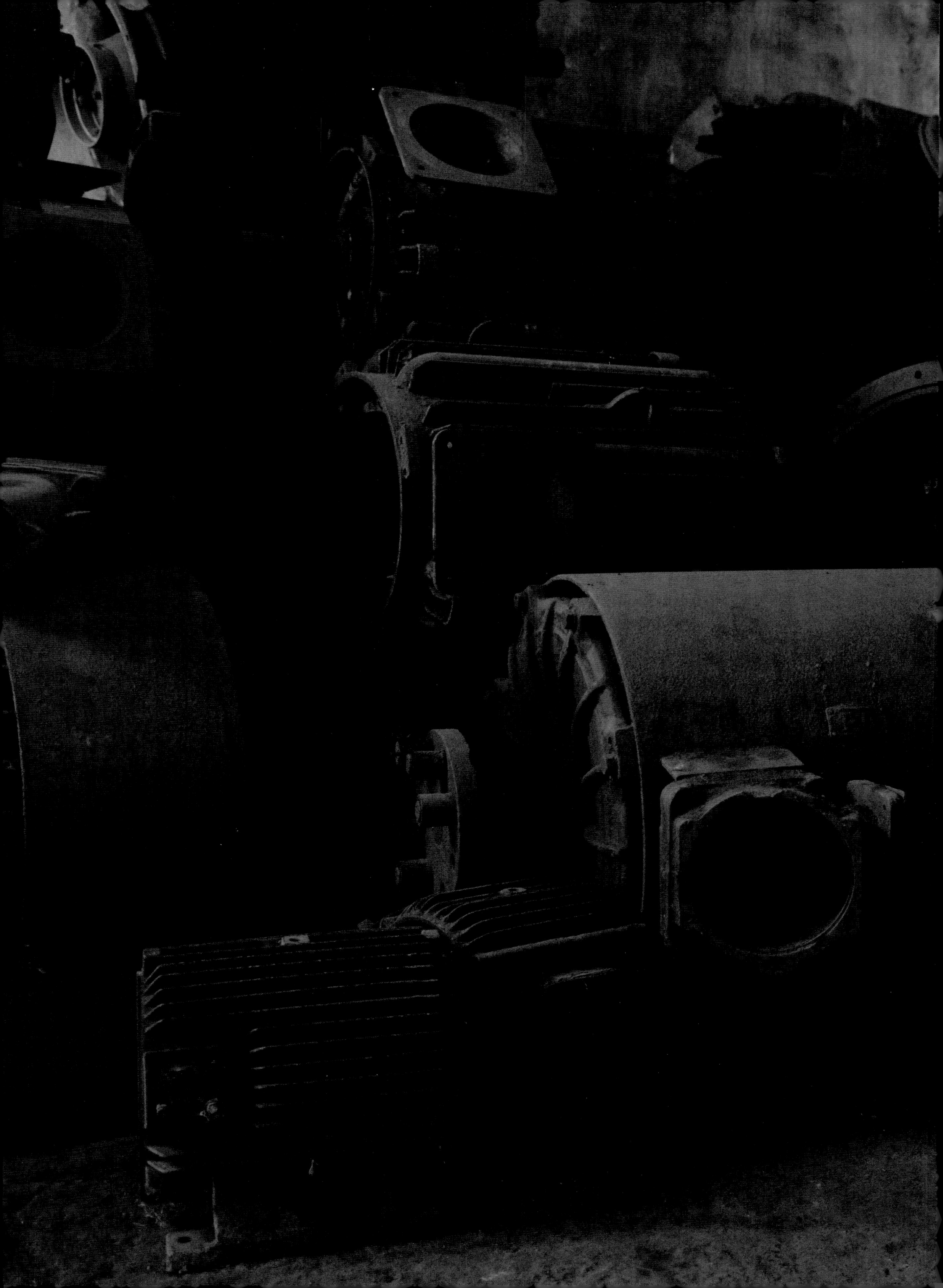

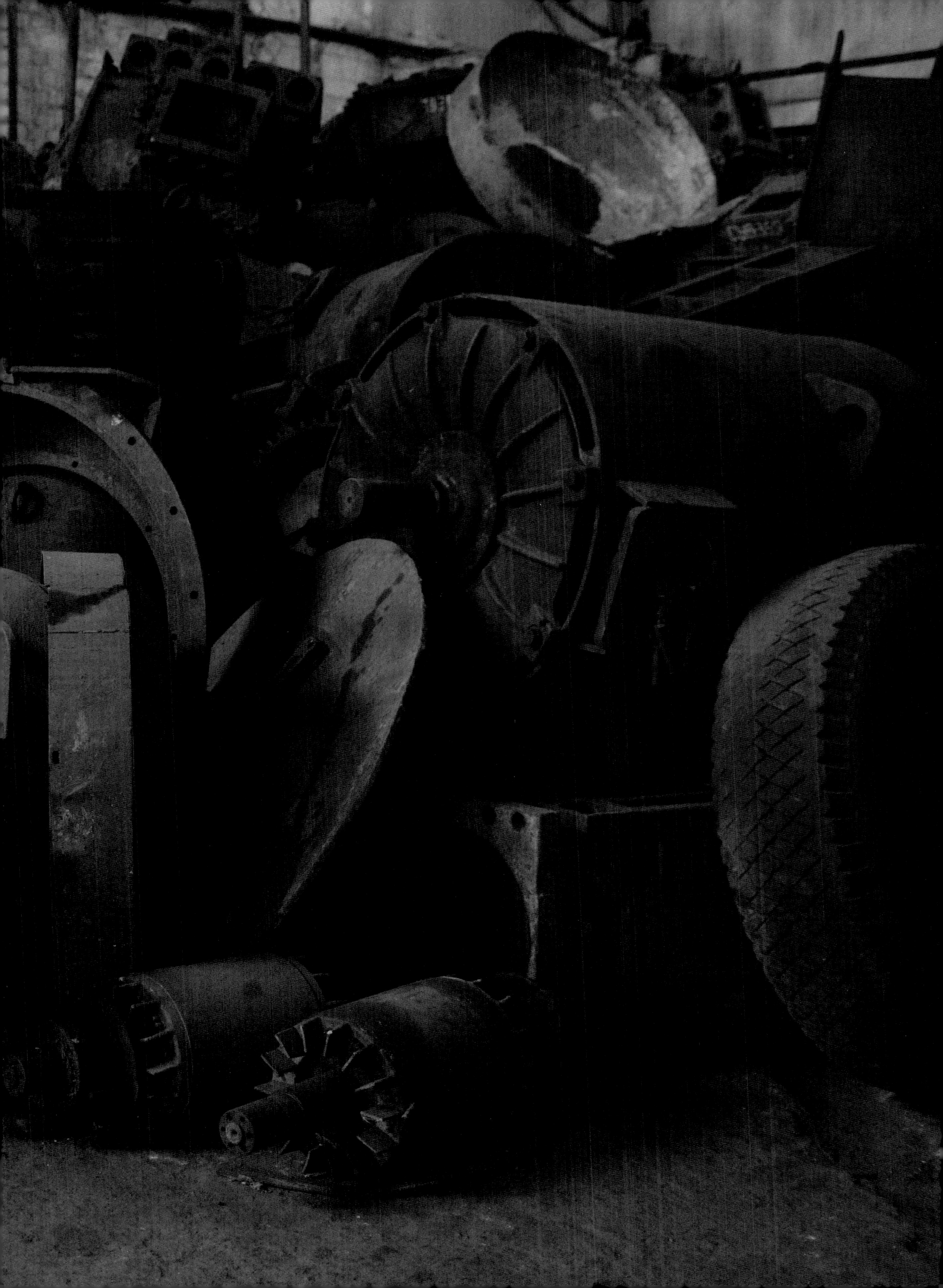

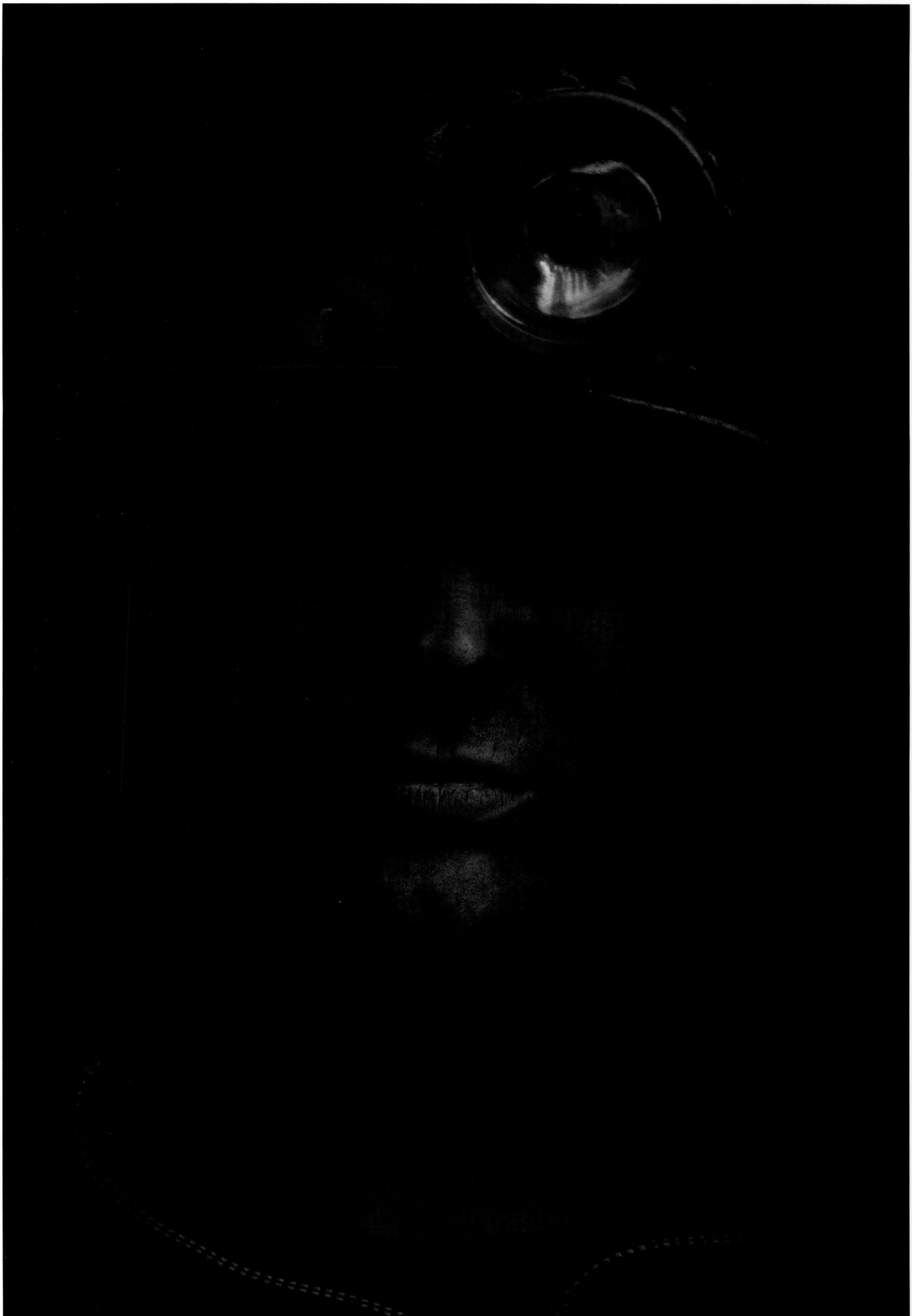

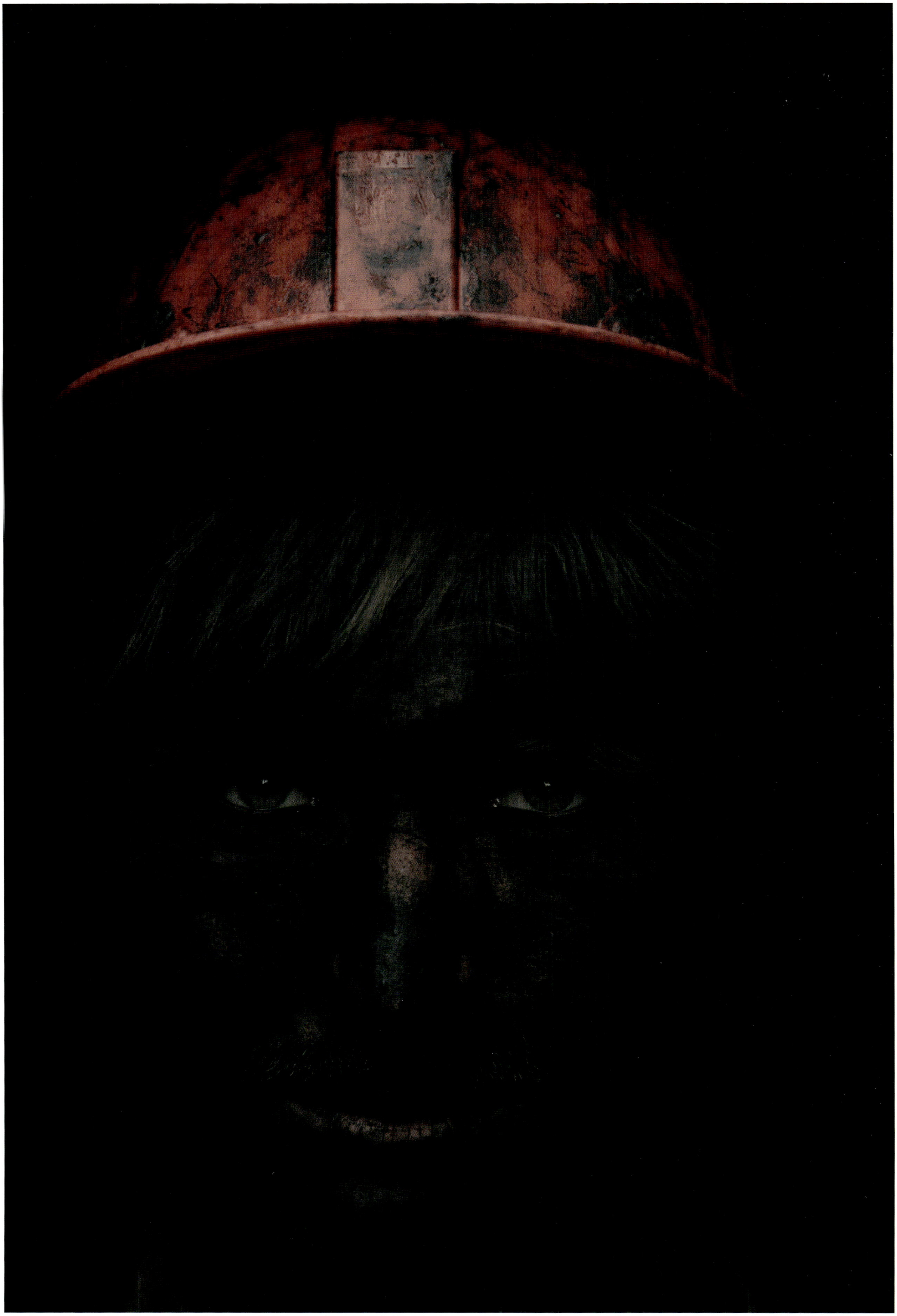

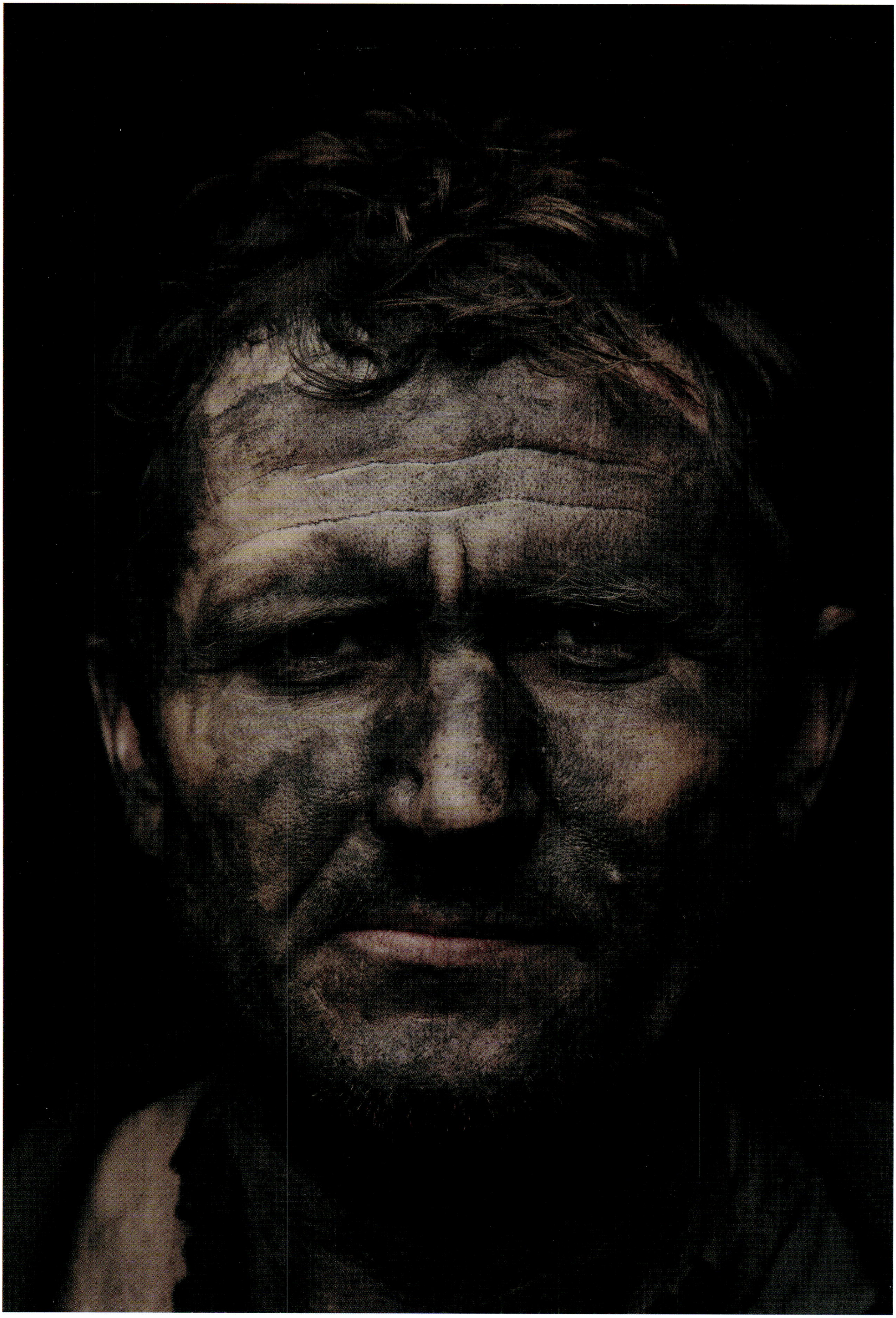

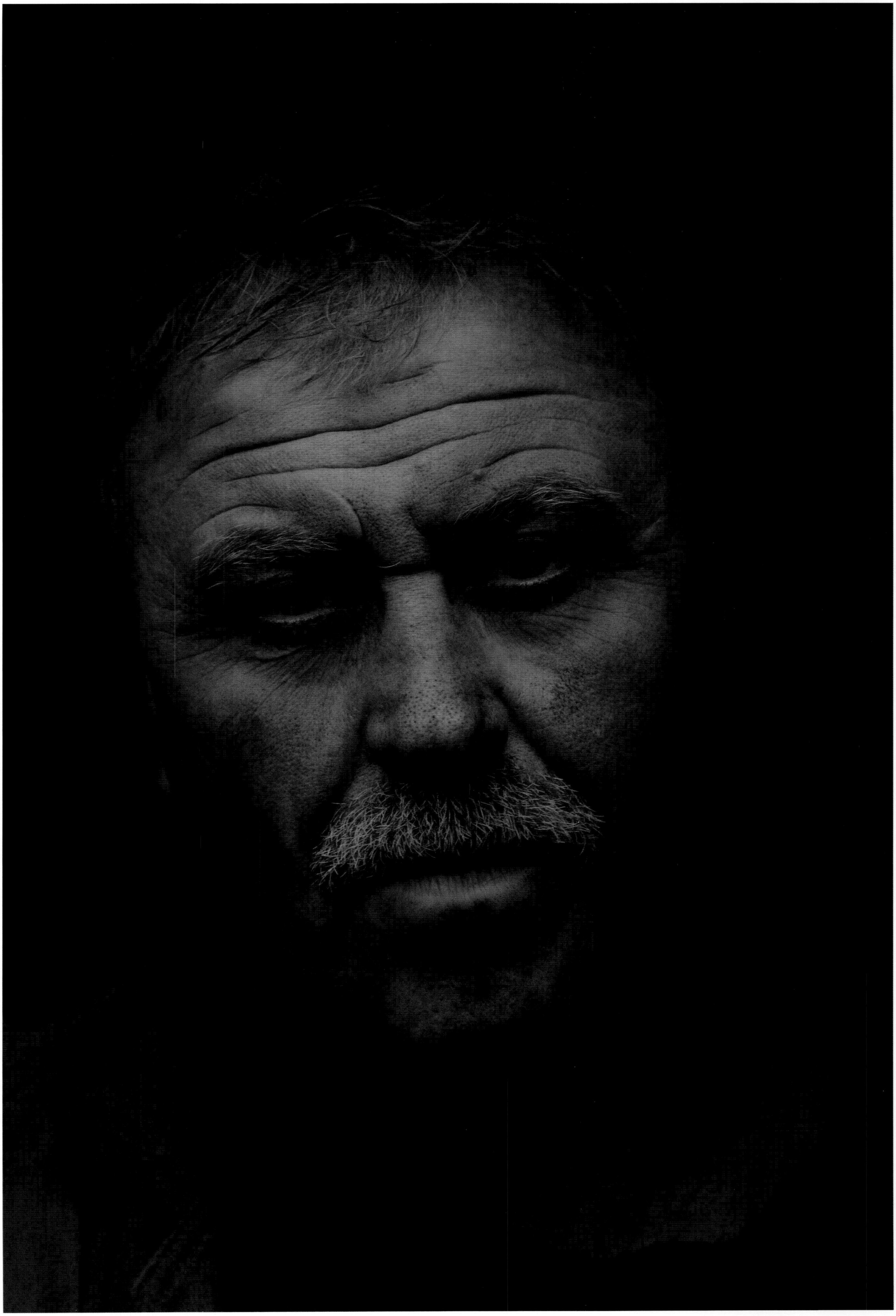

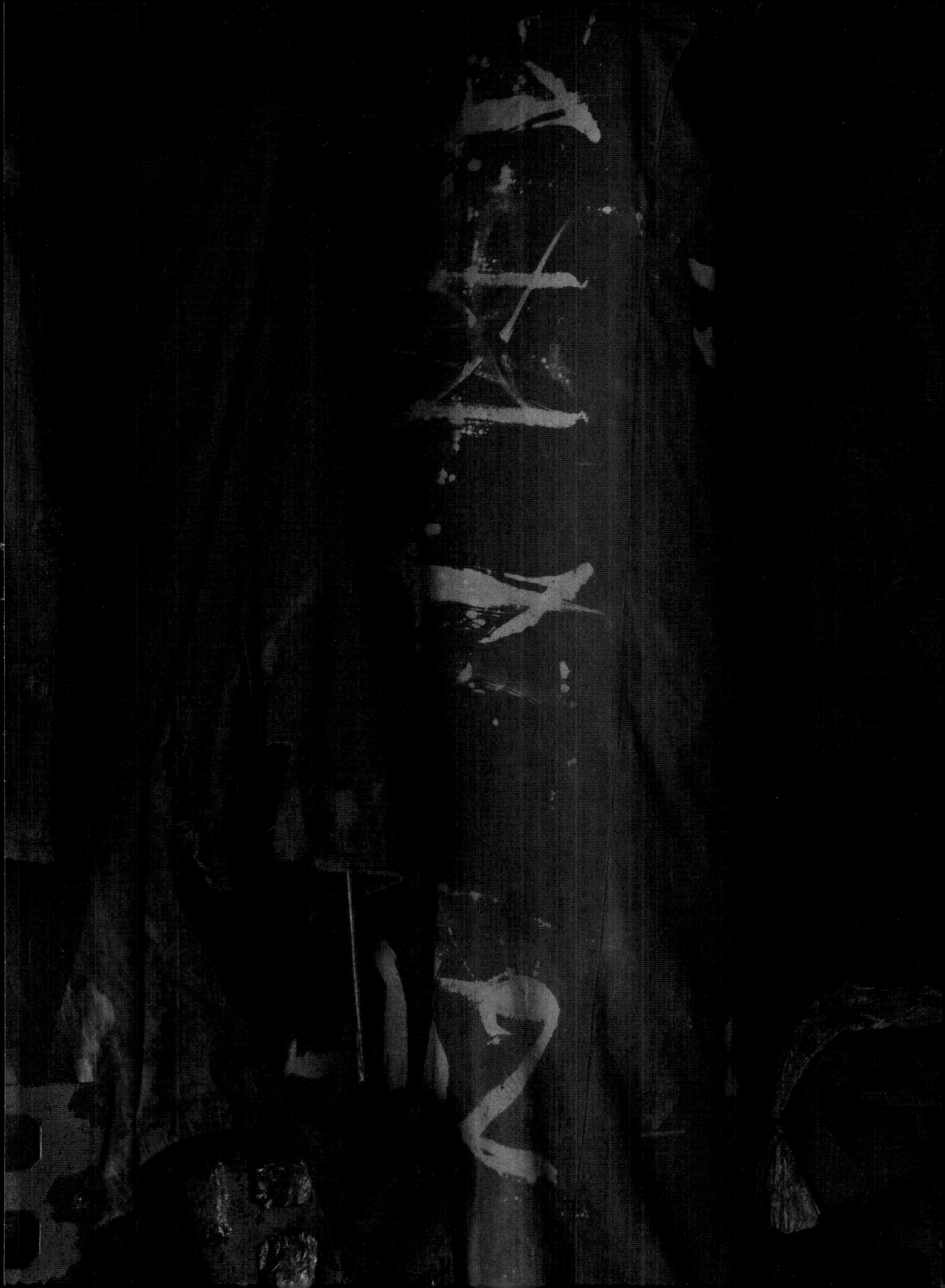

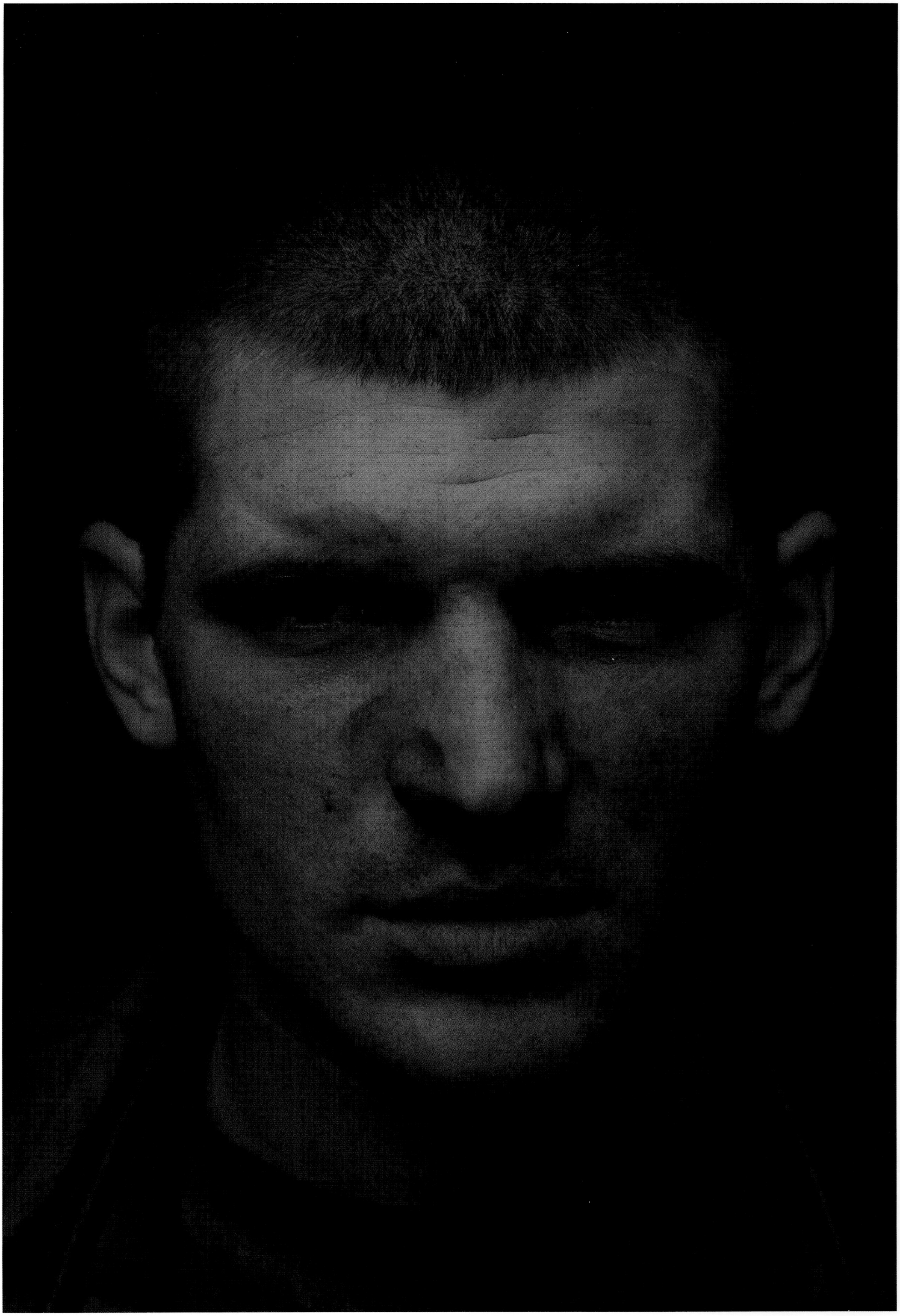

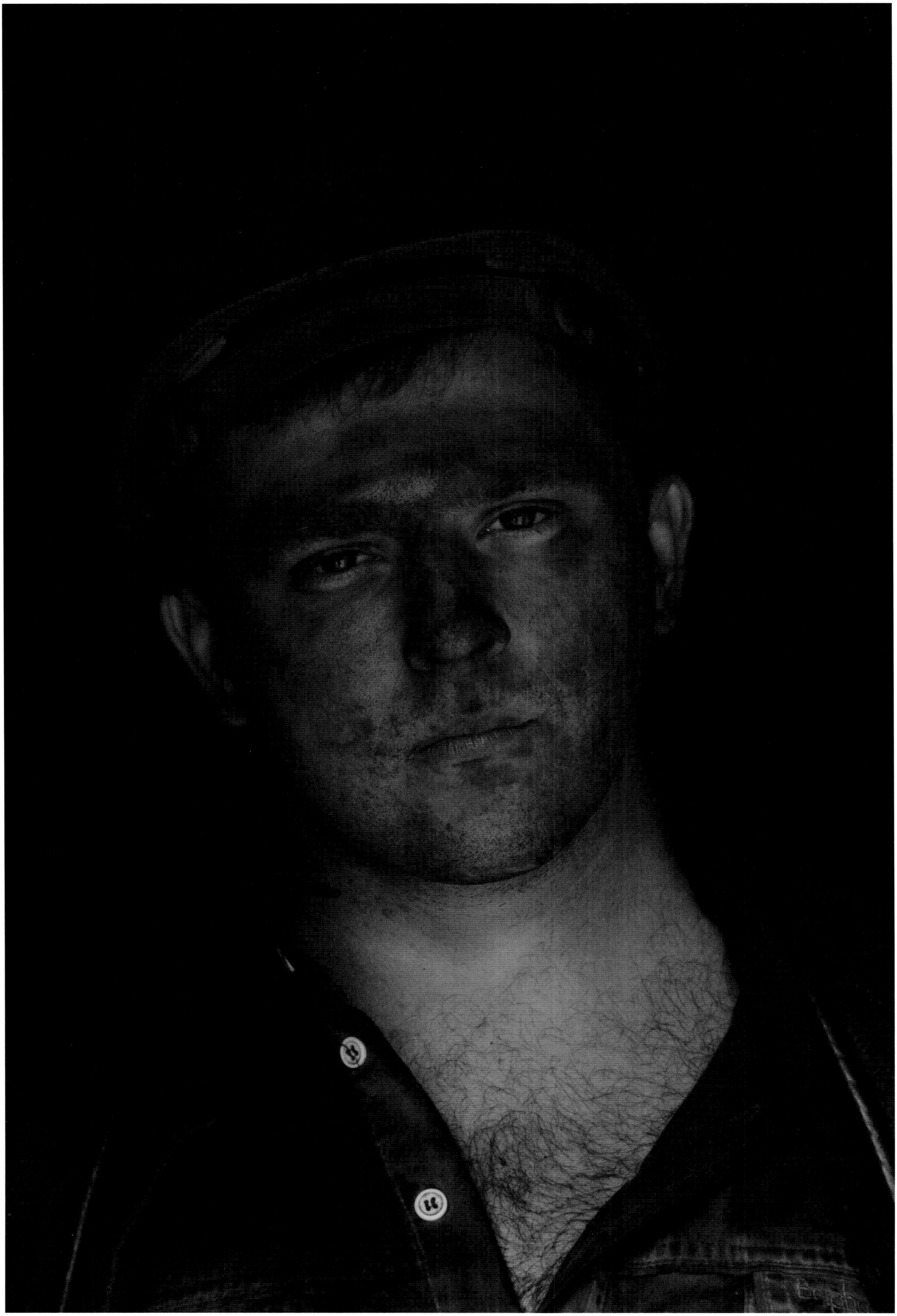

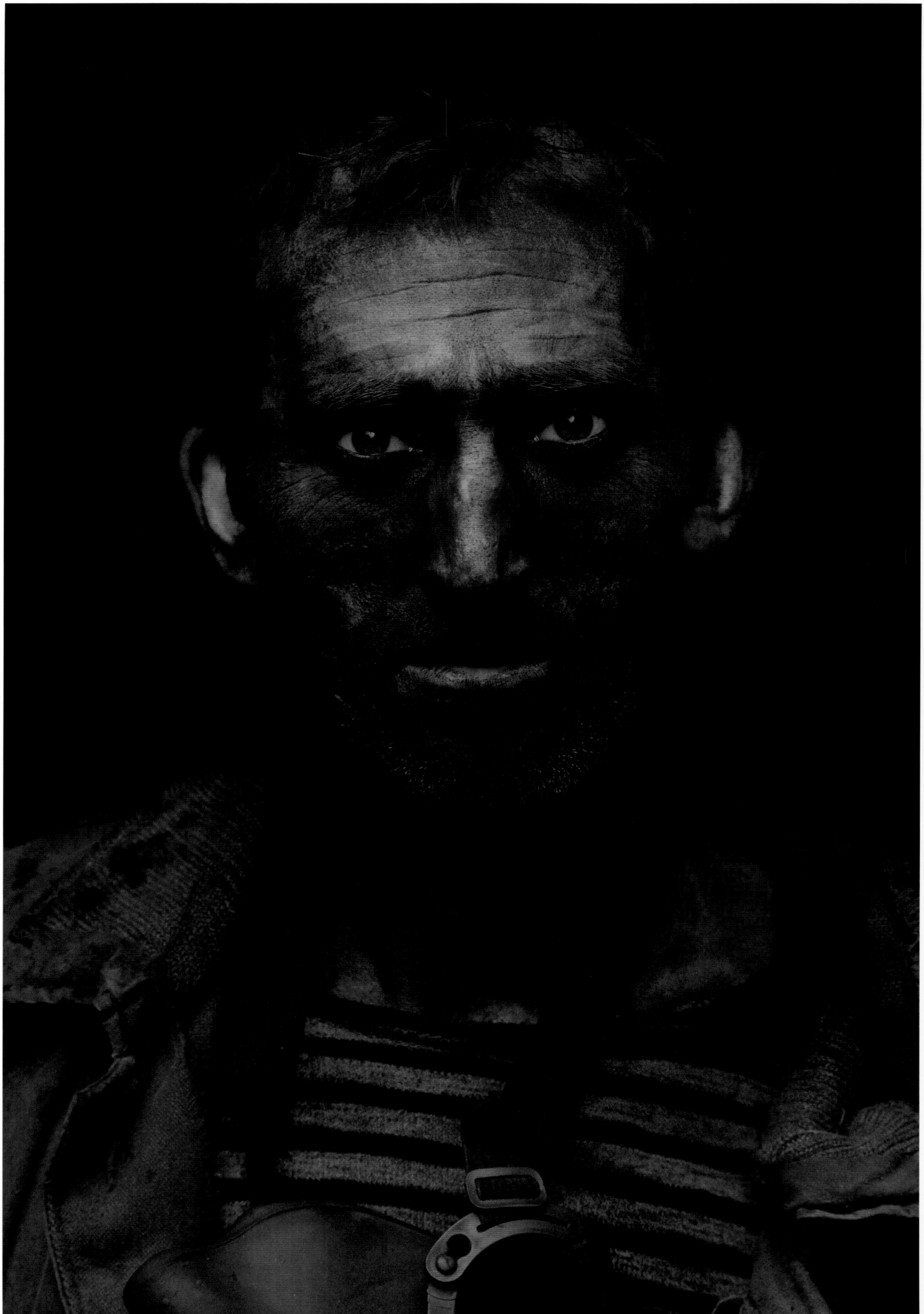

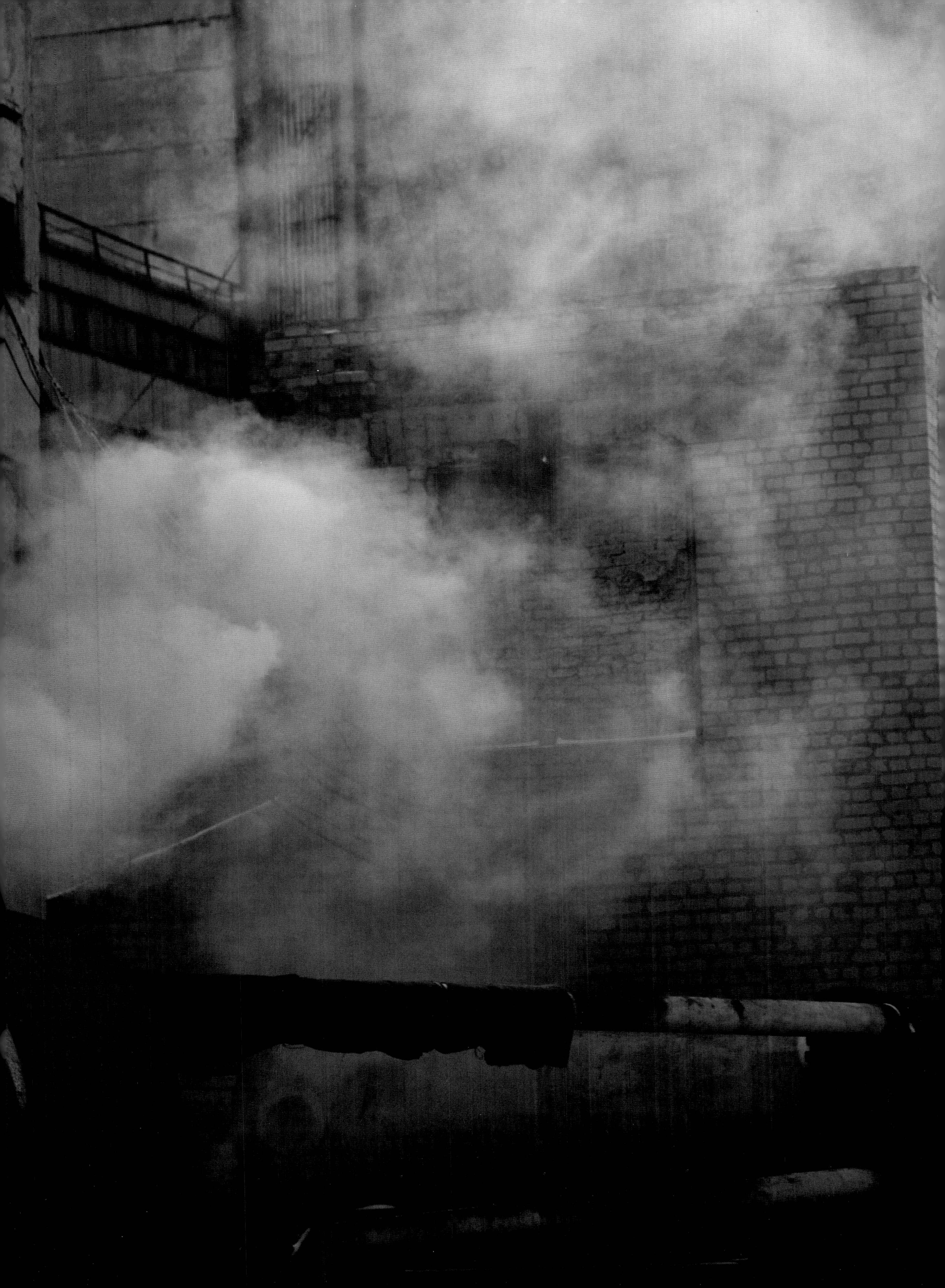

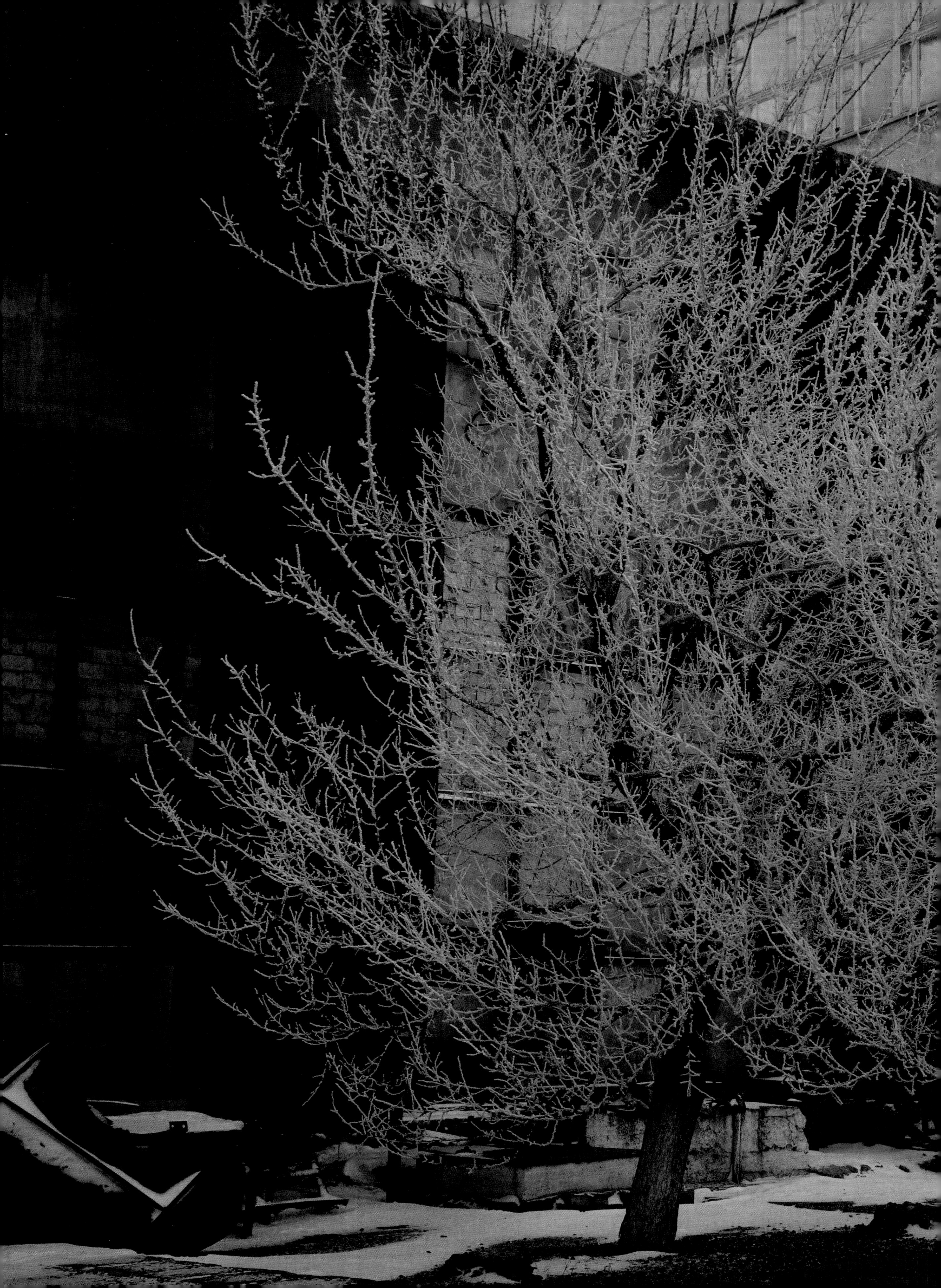

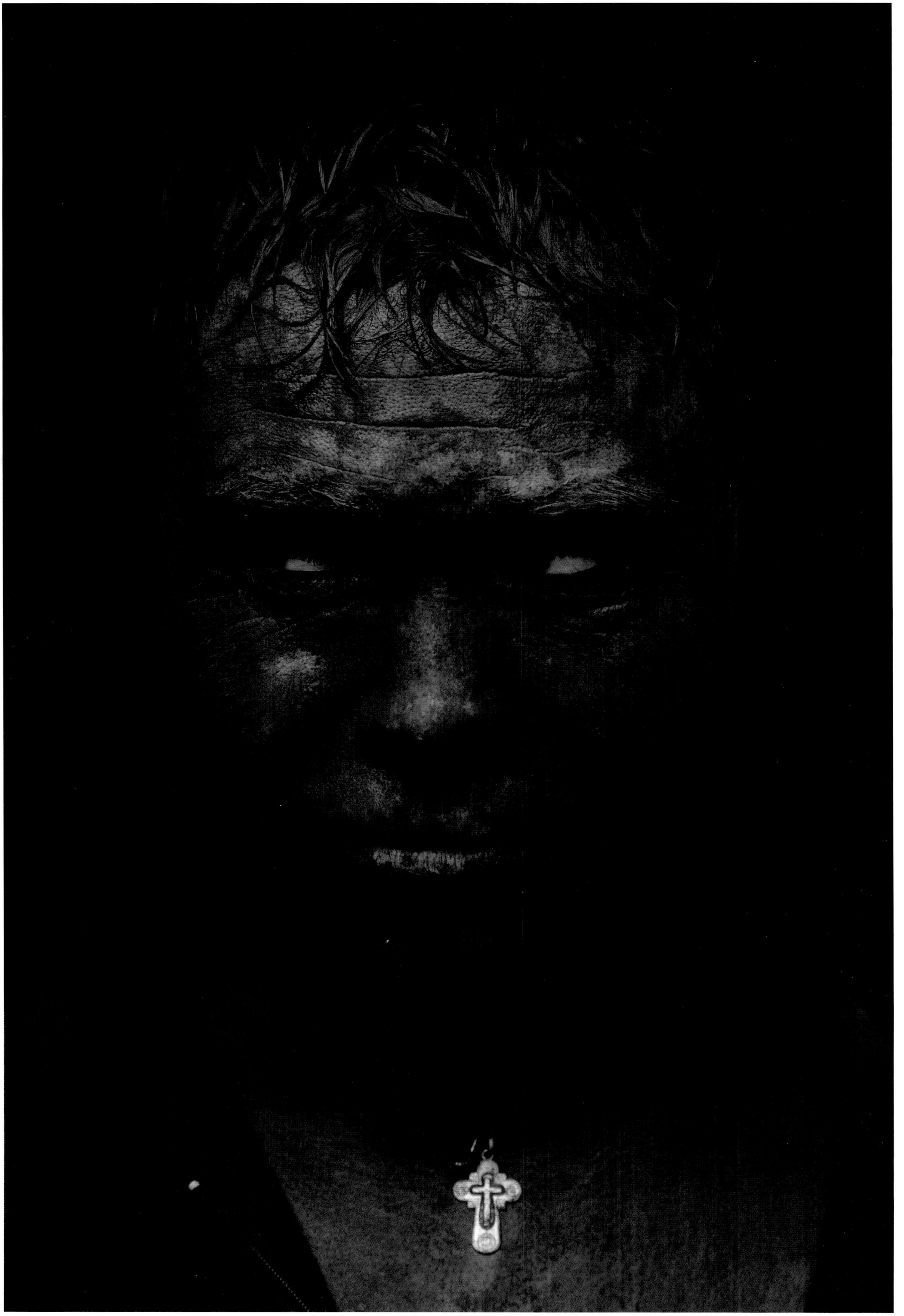

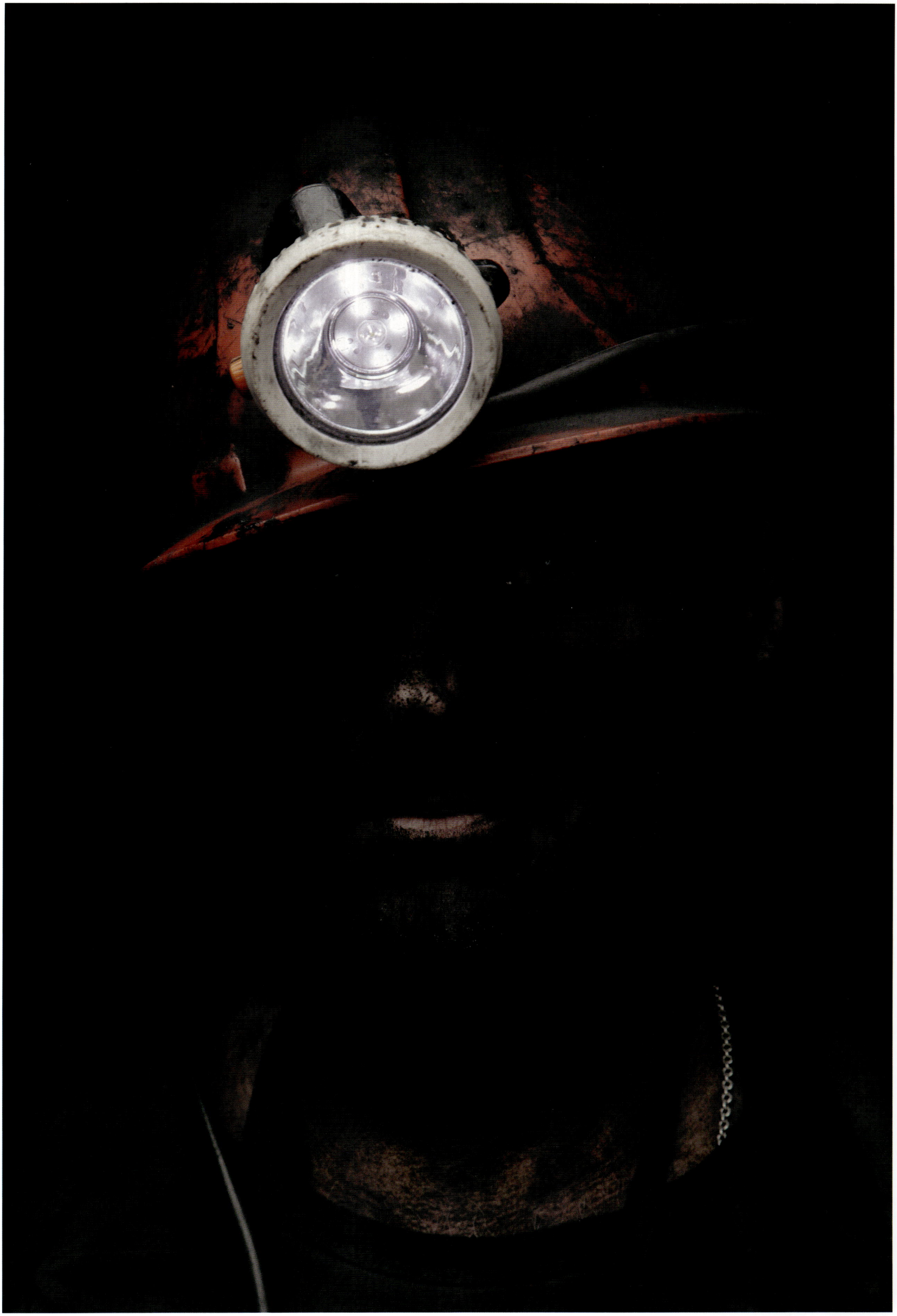

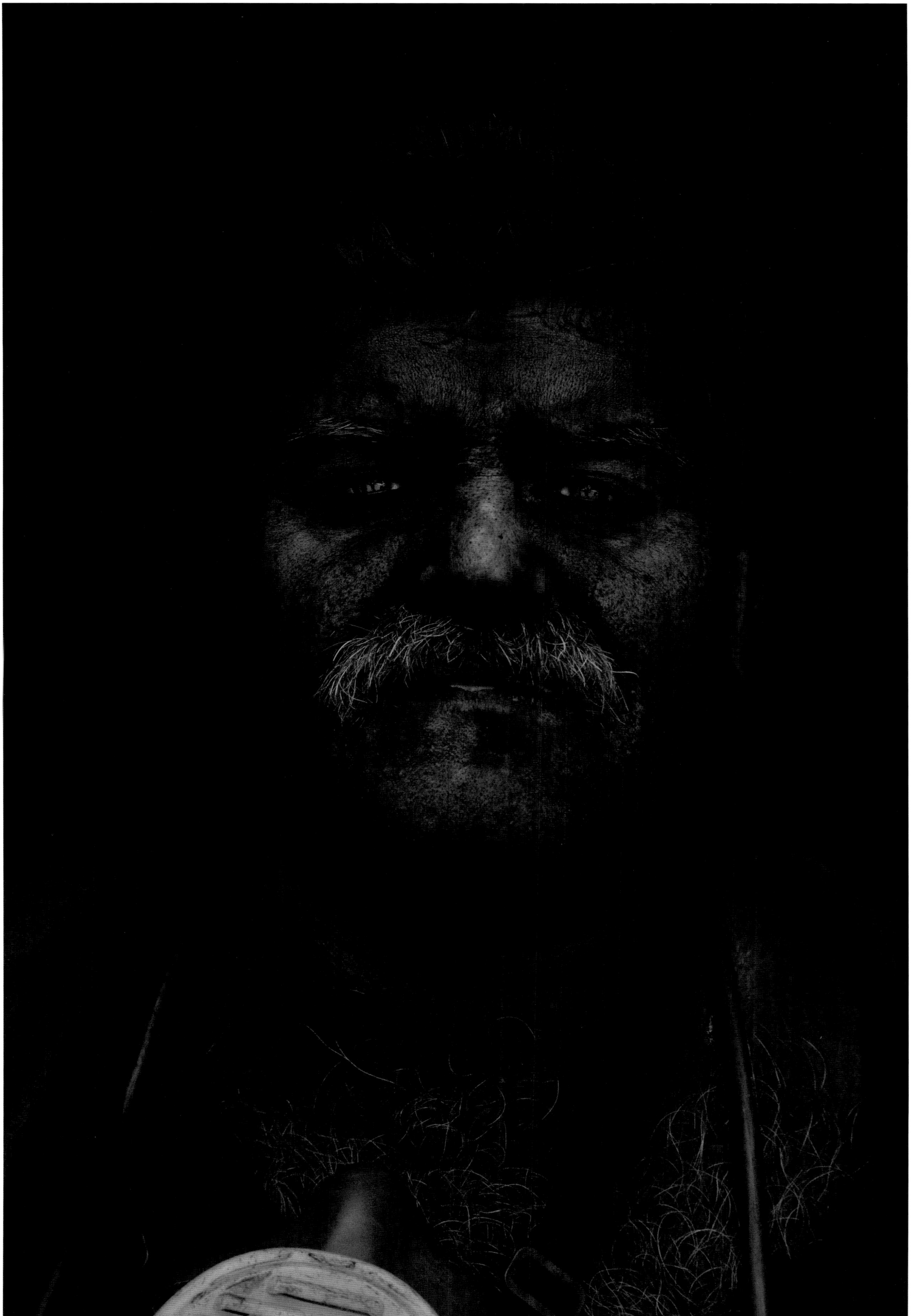

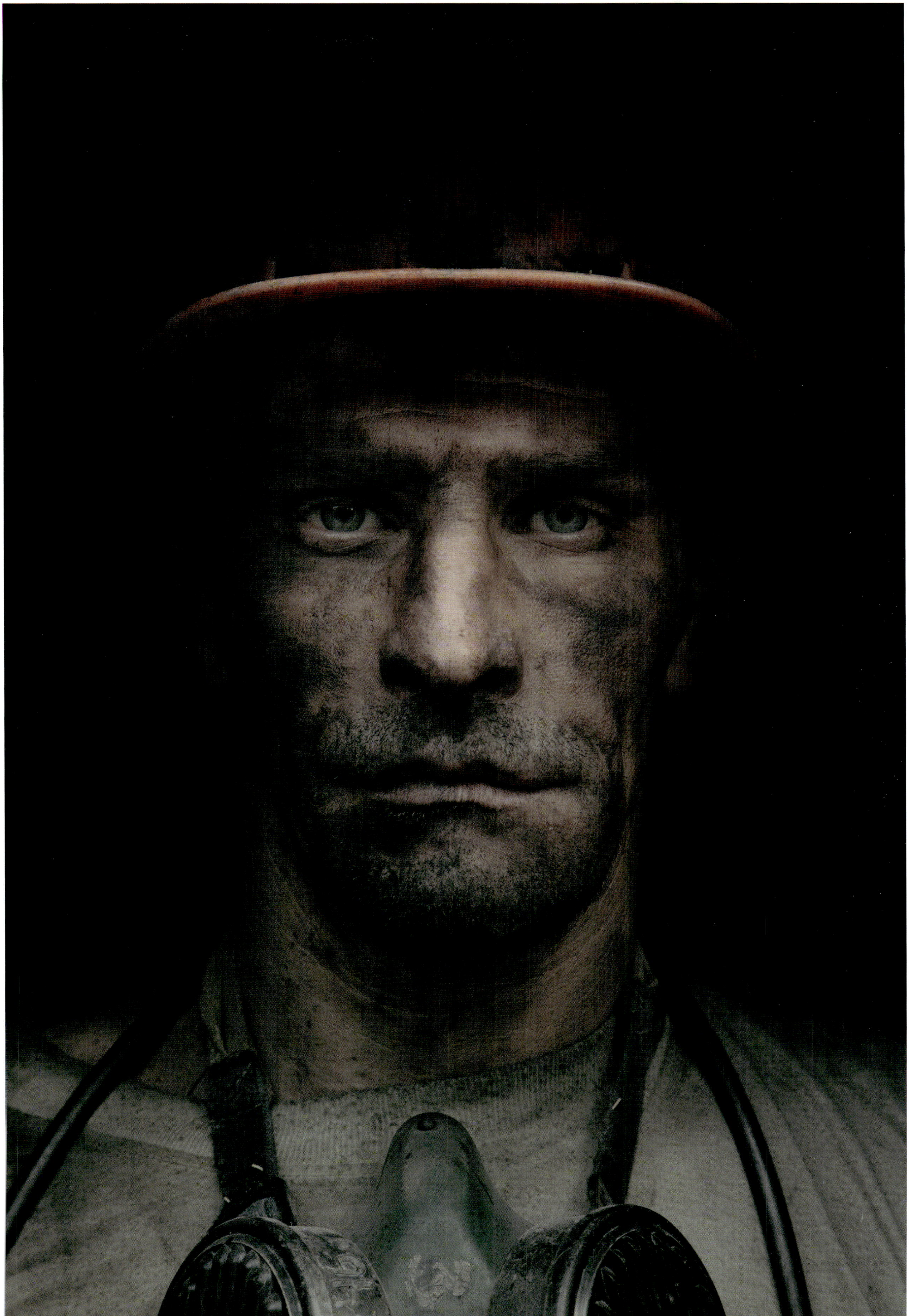

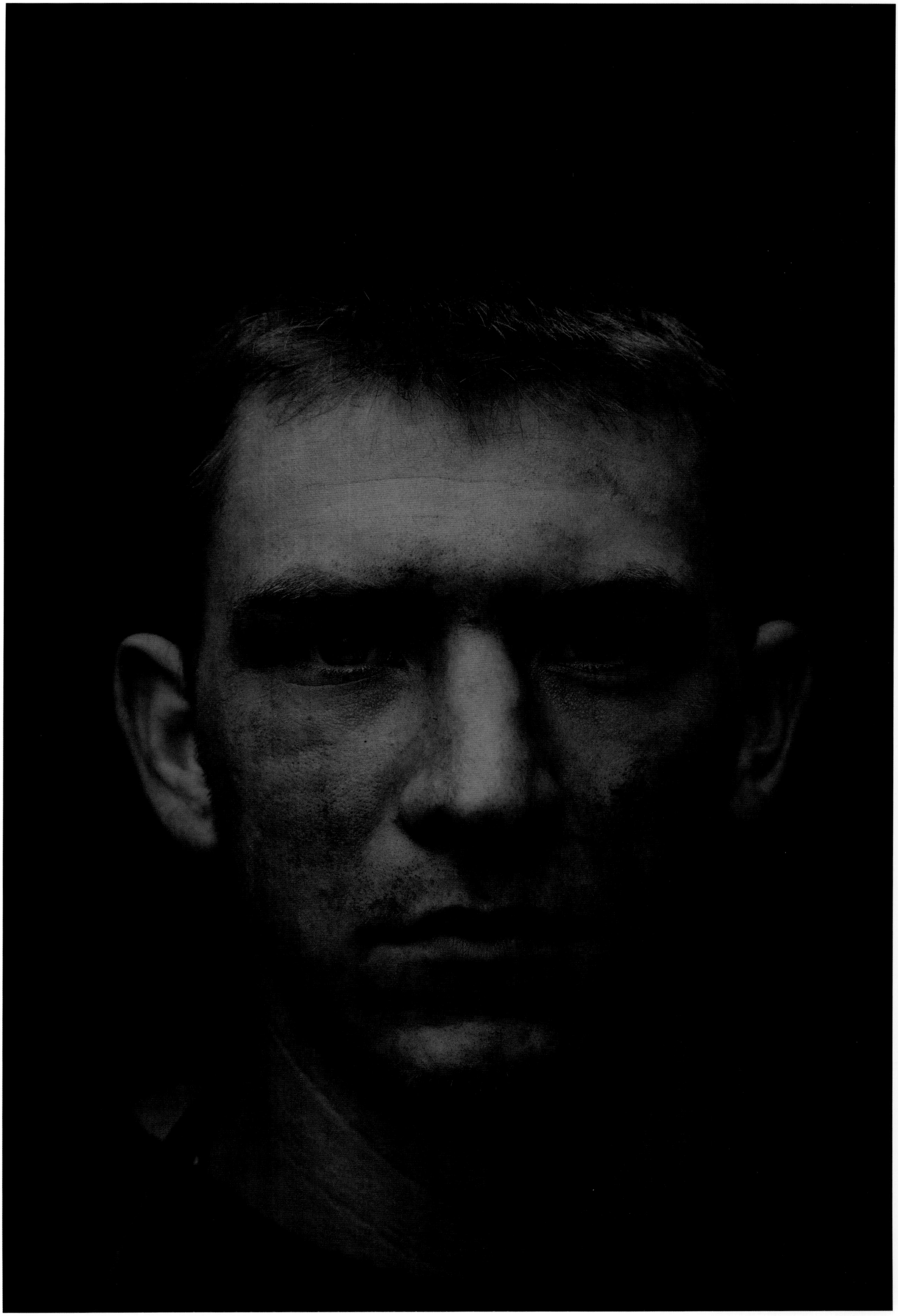

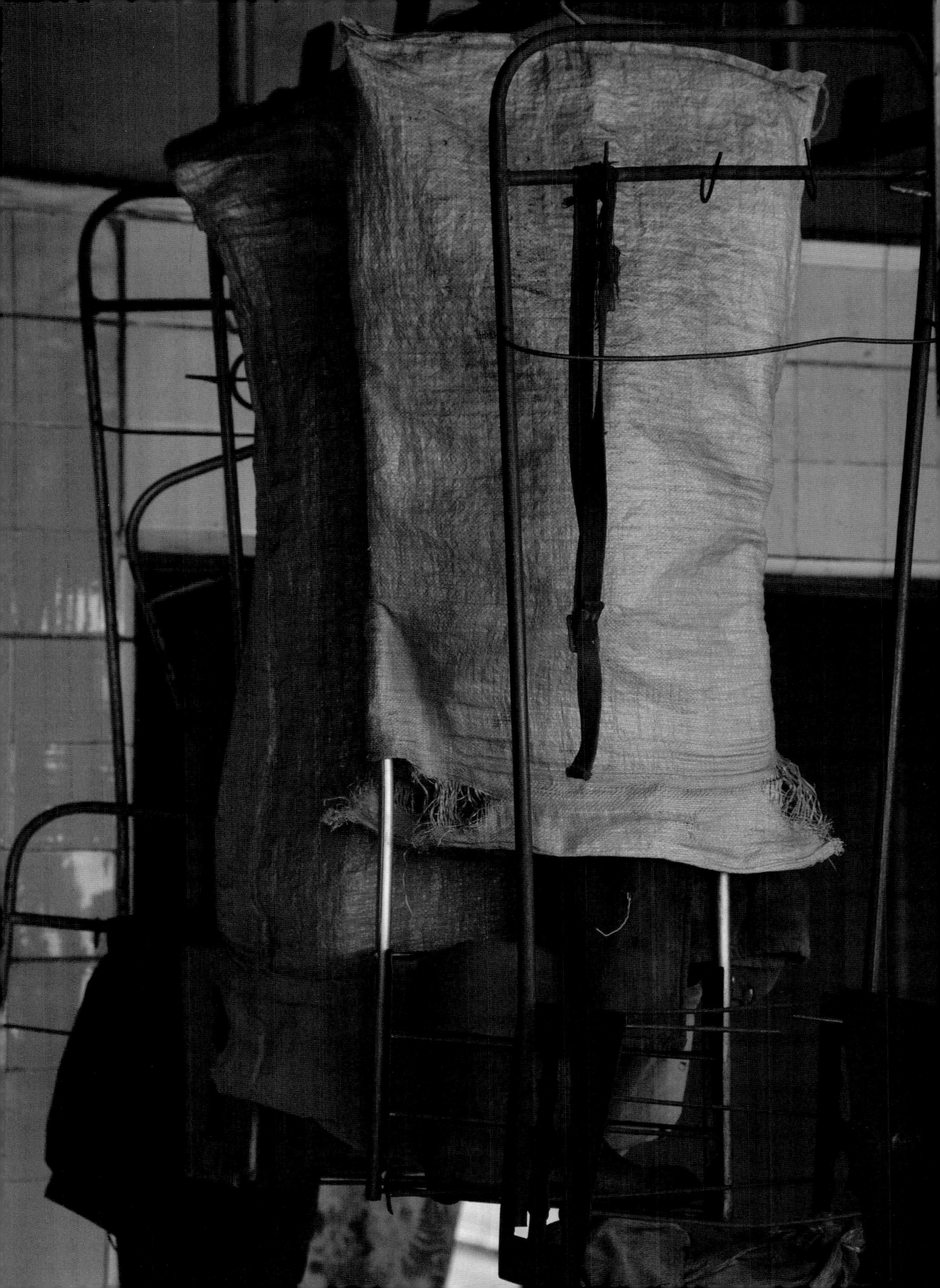

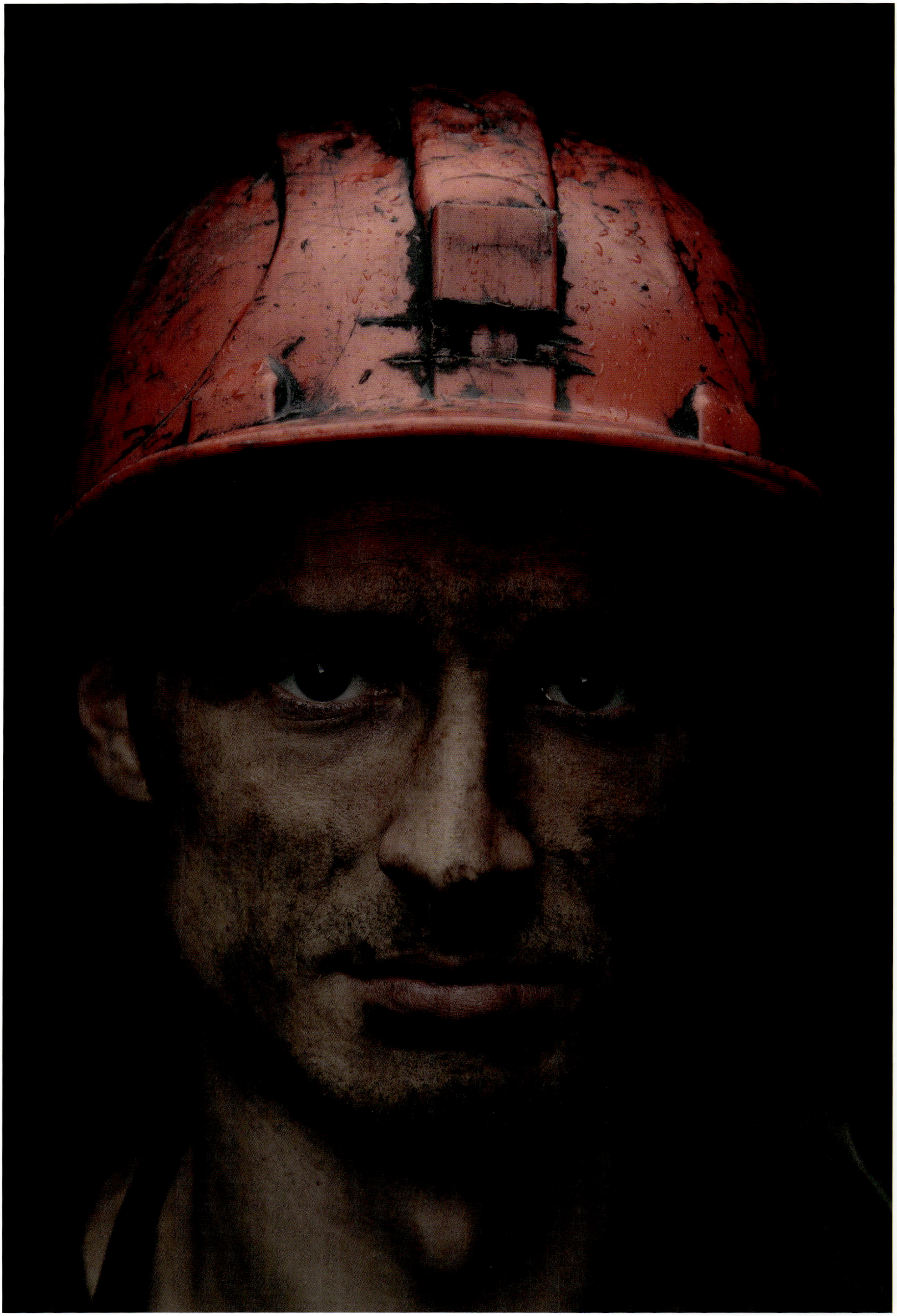

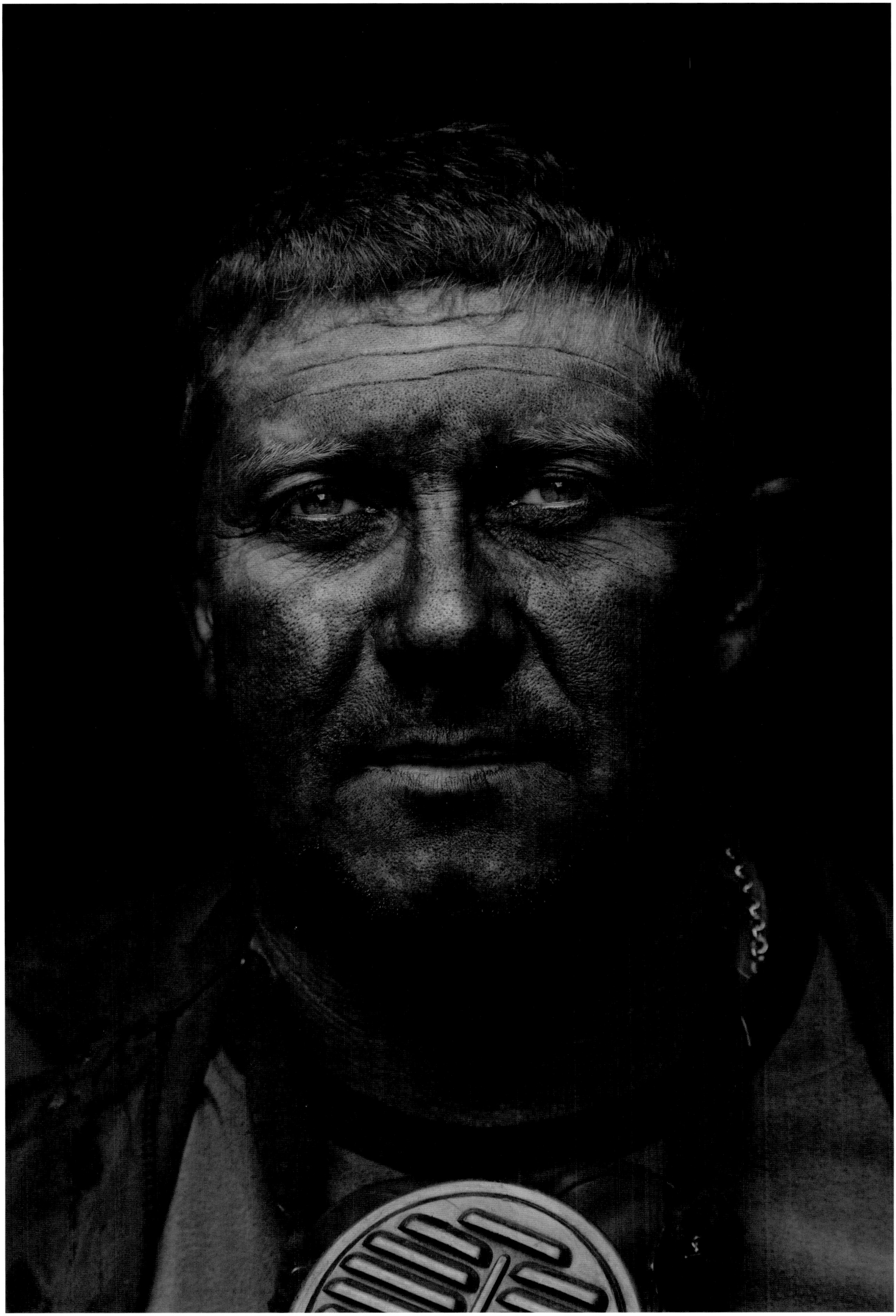

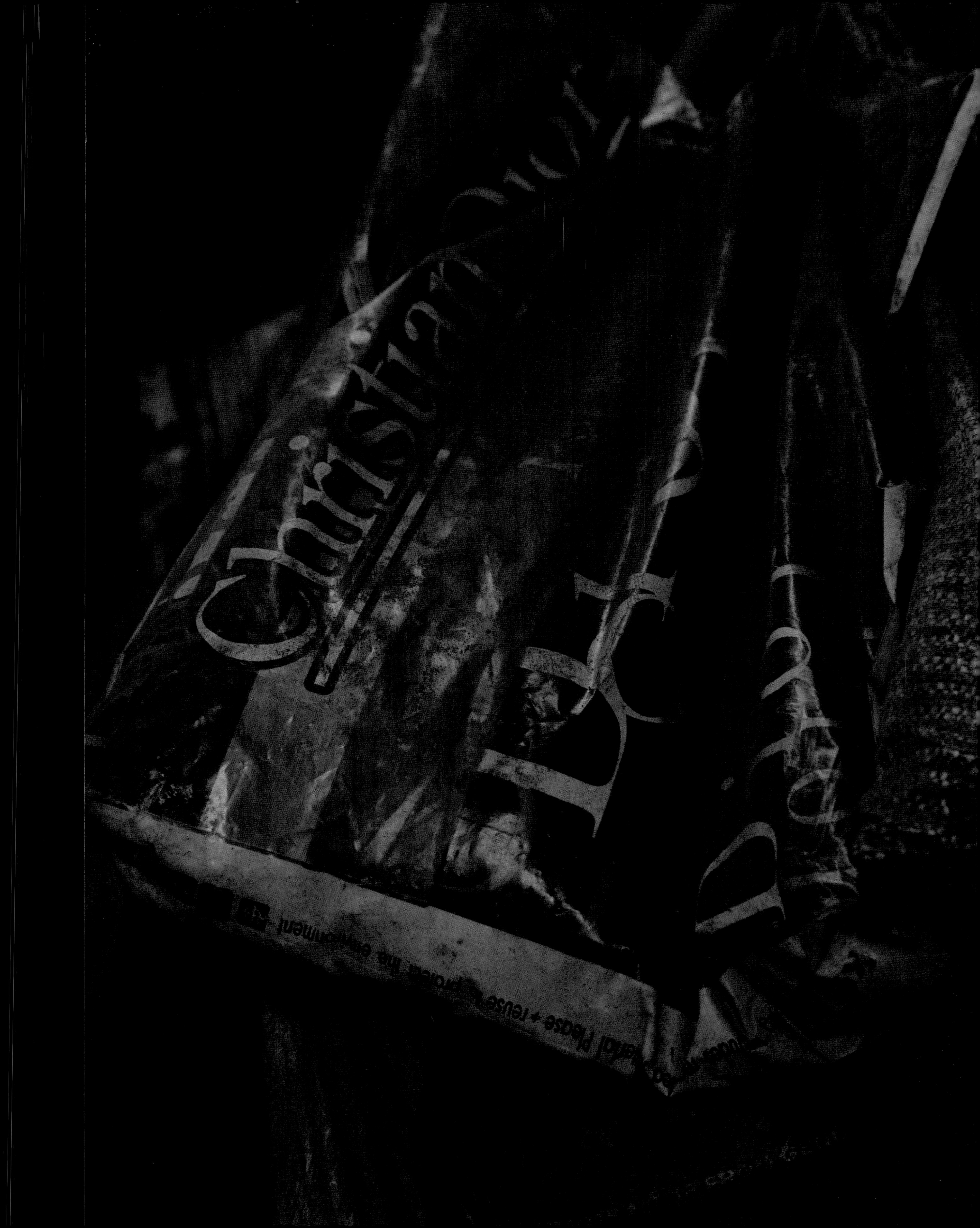
Christian

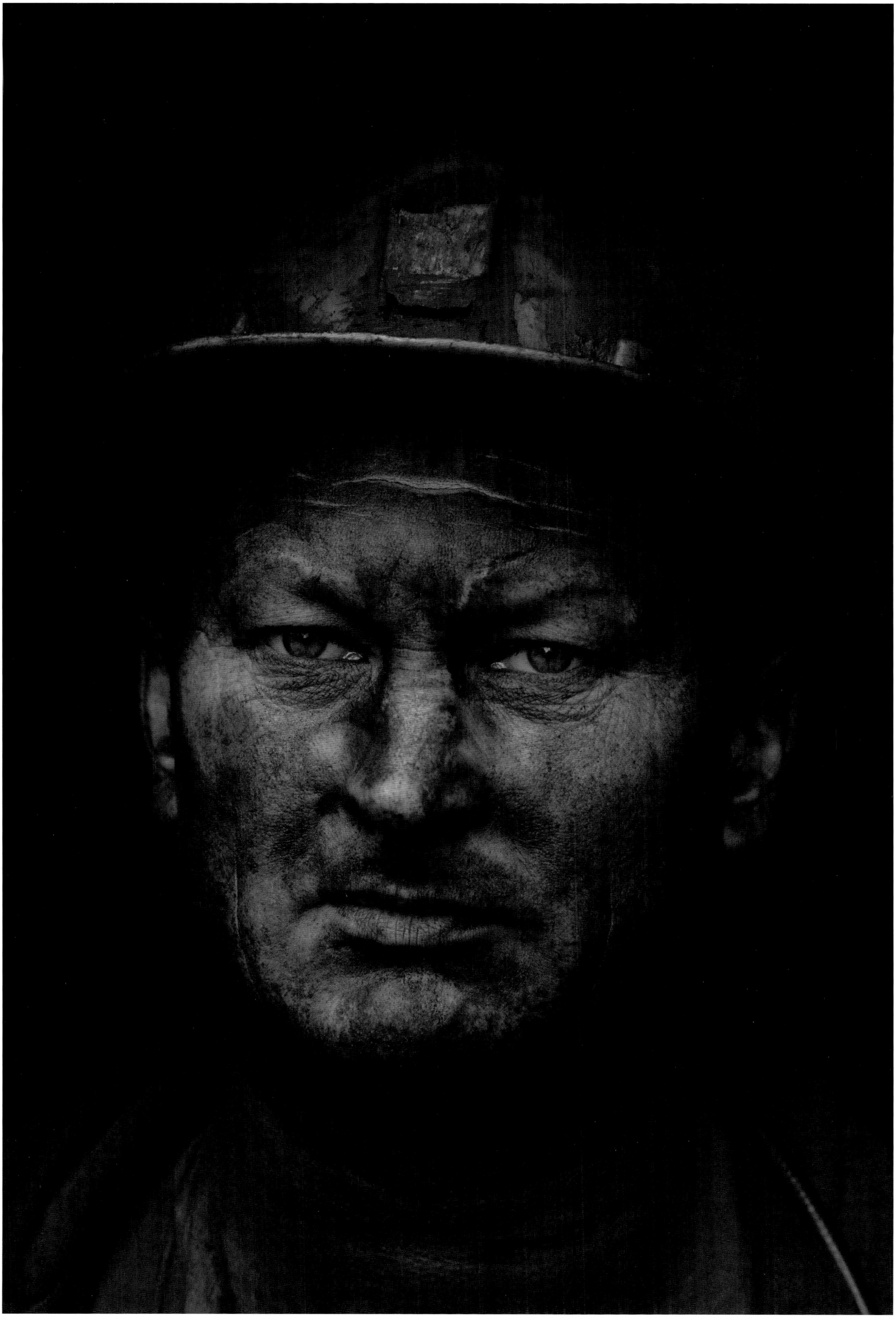

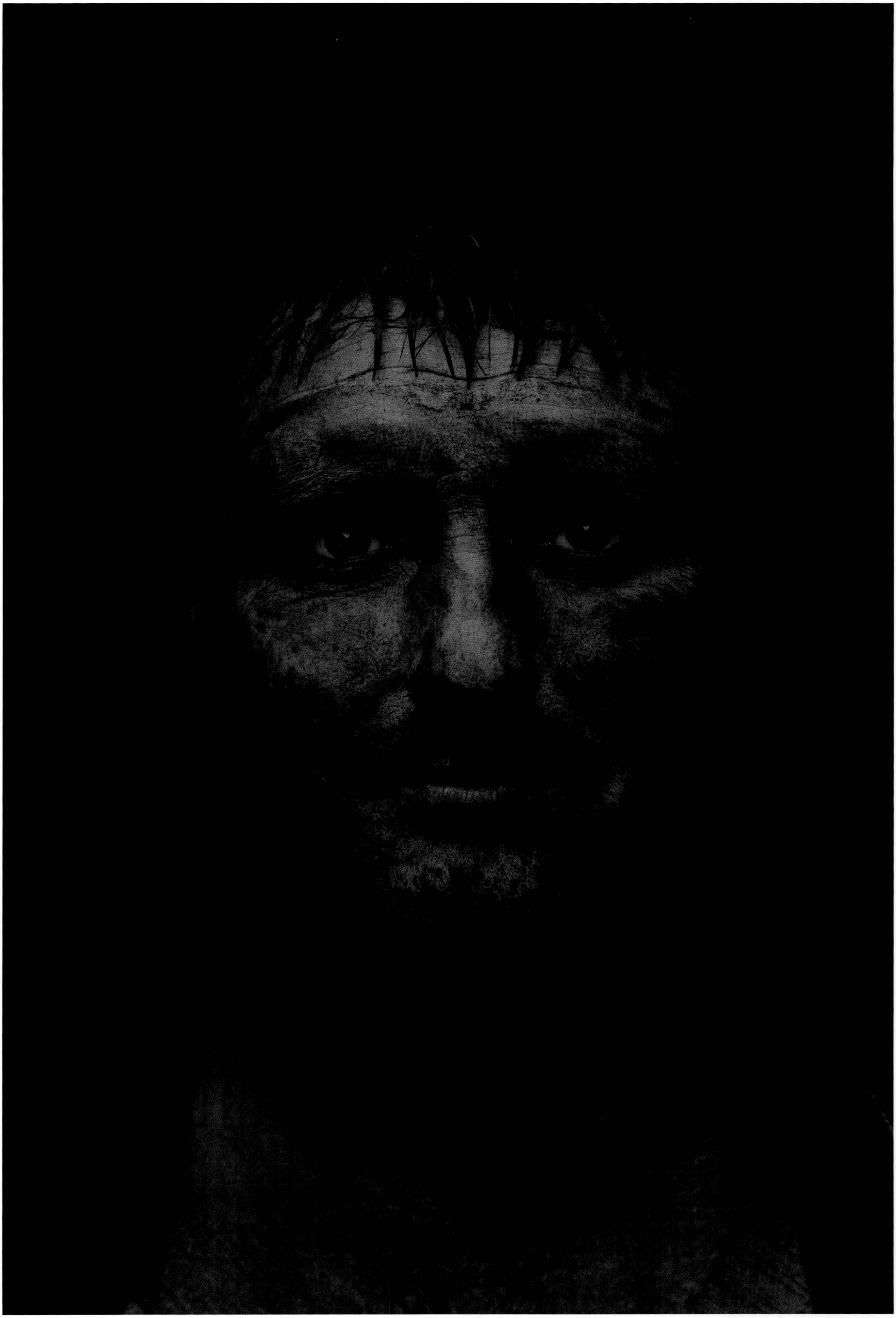

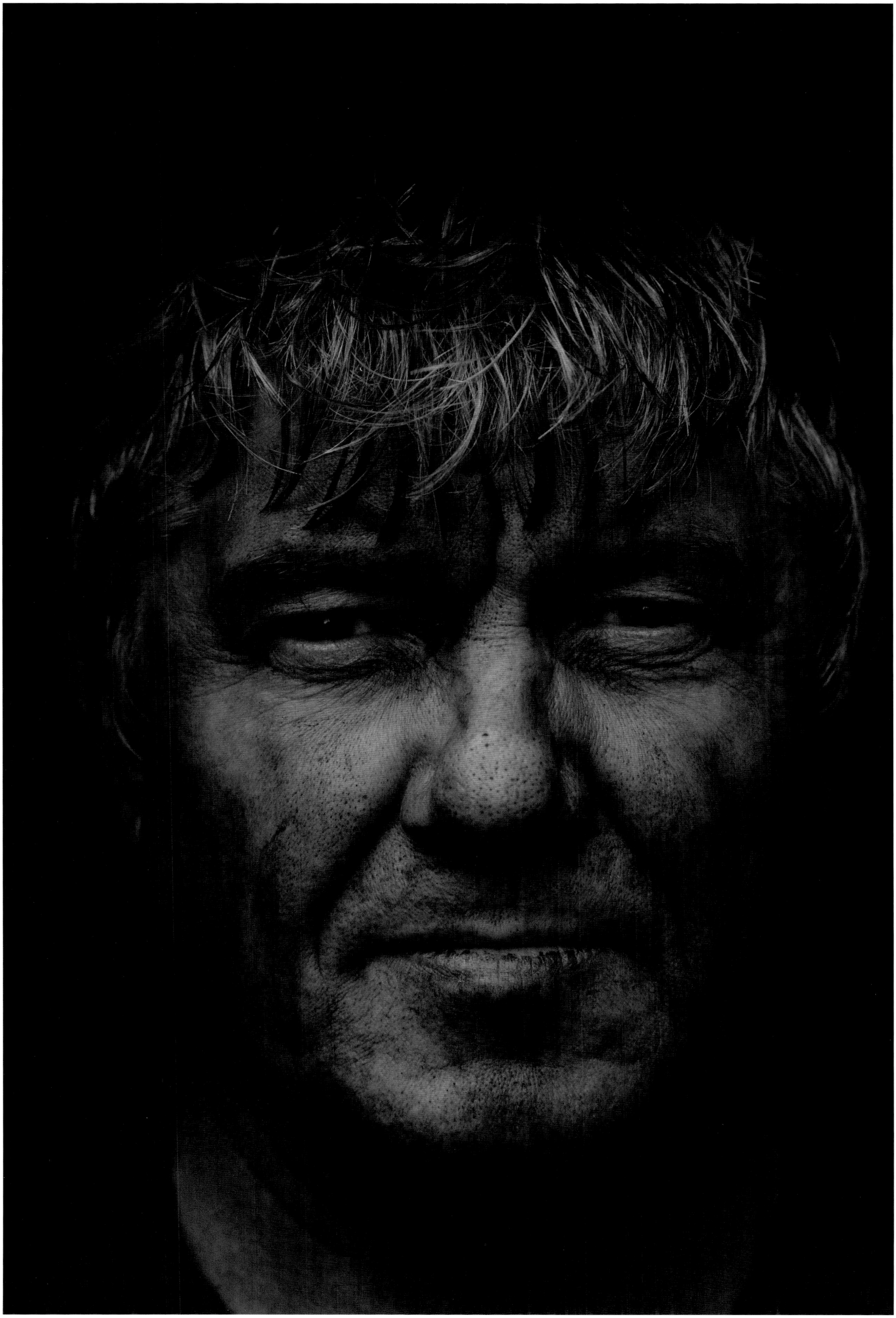

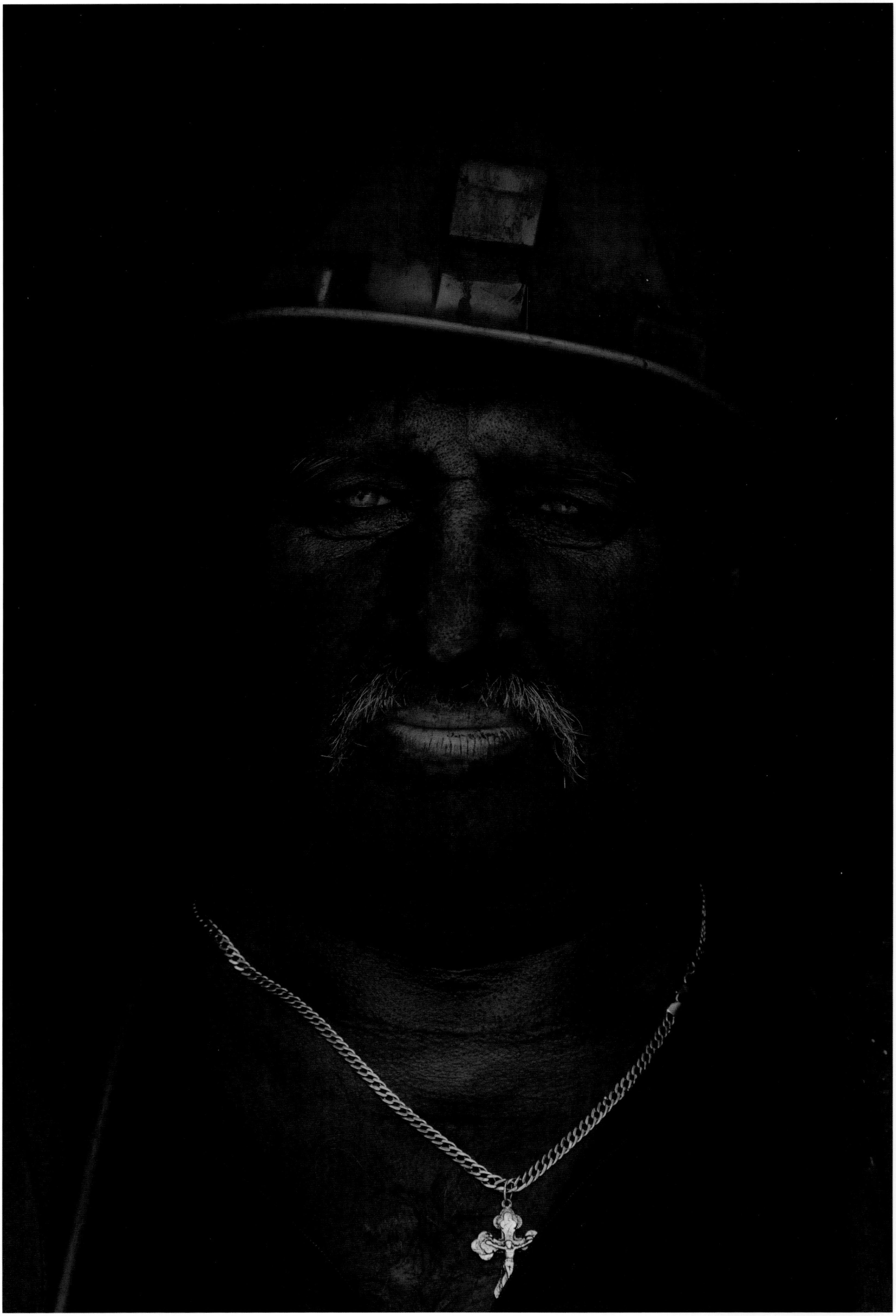

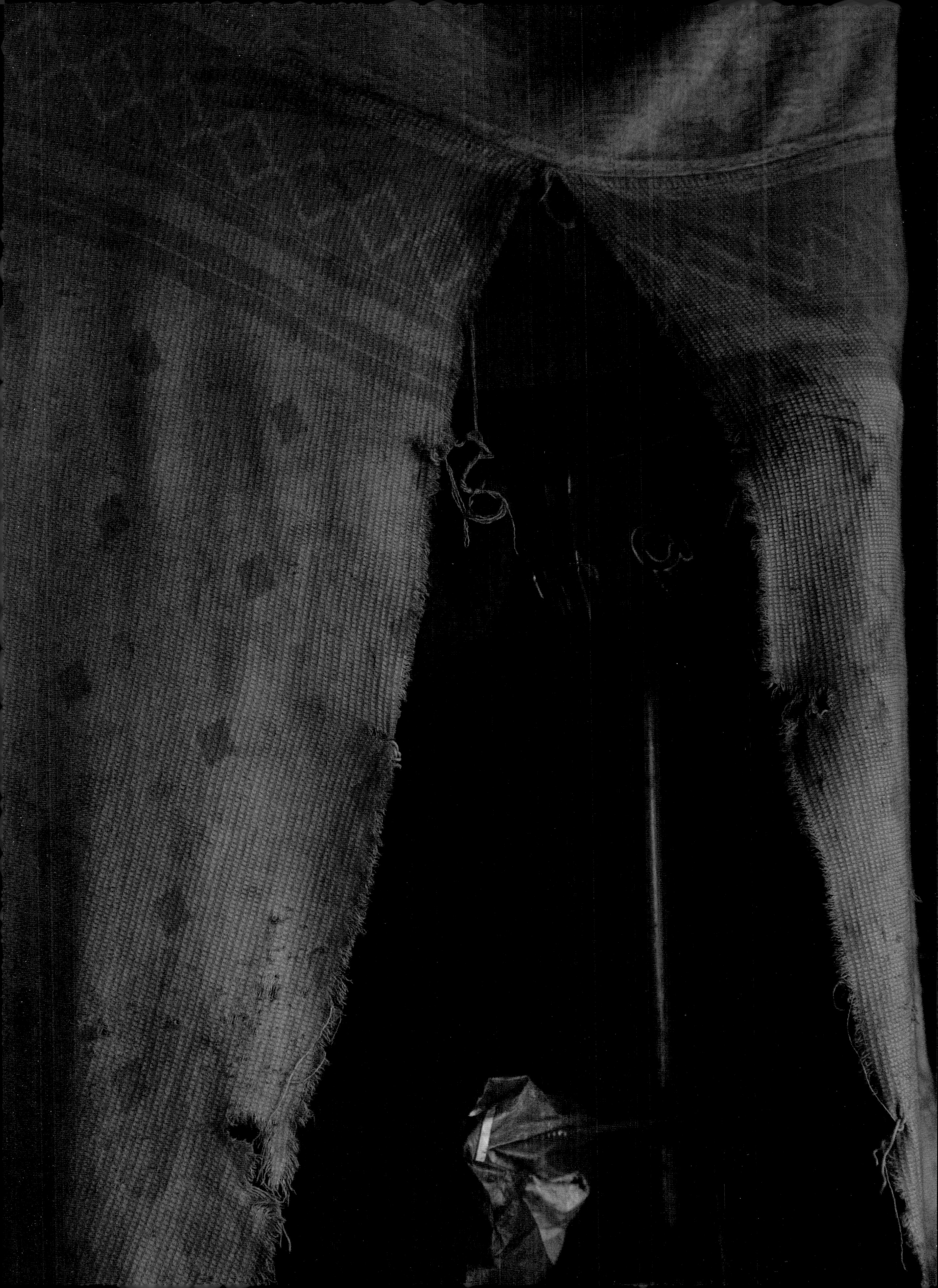

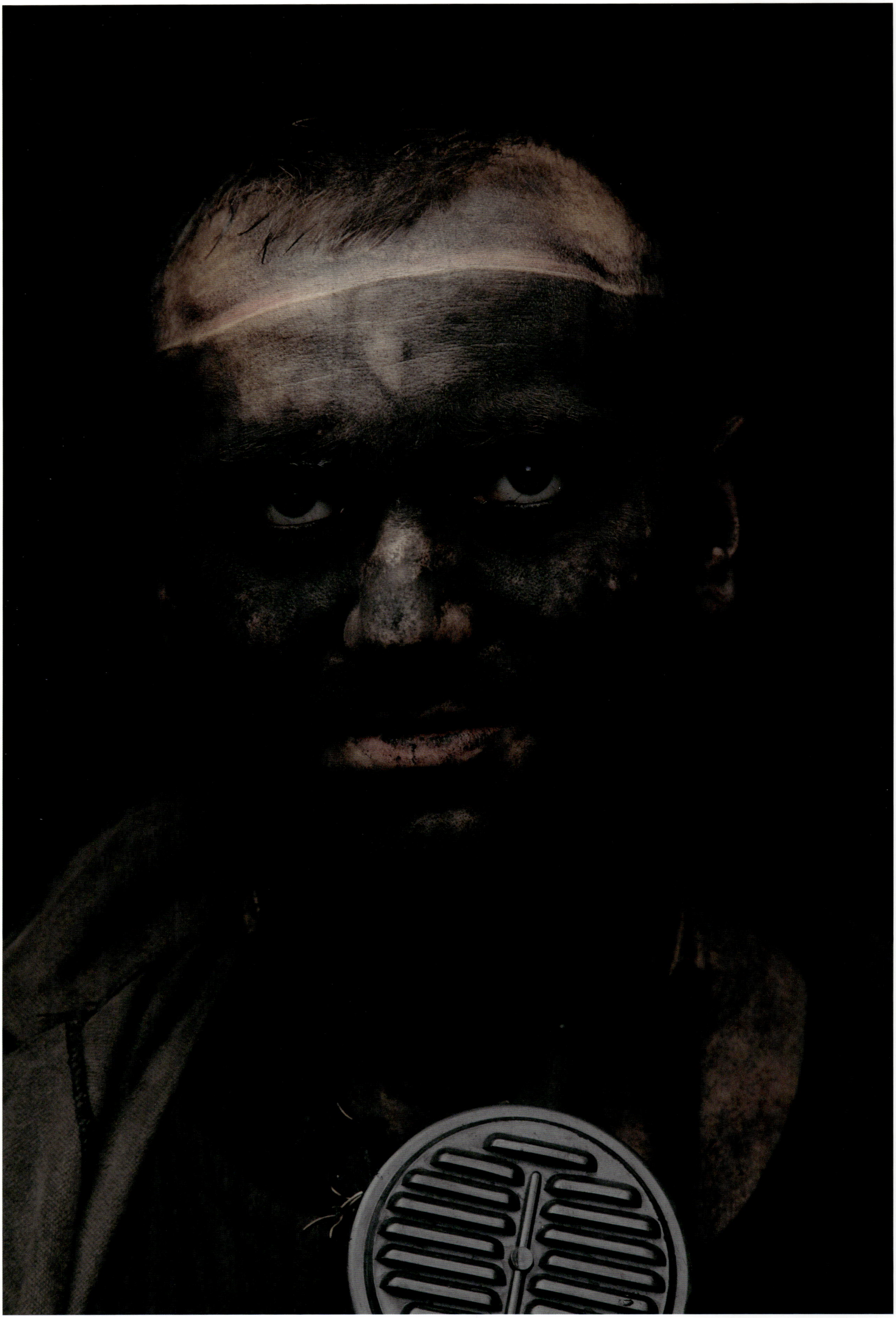

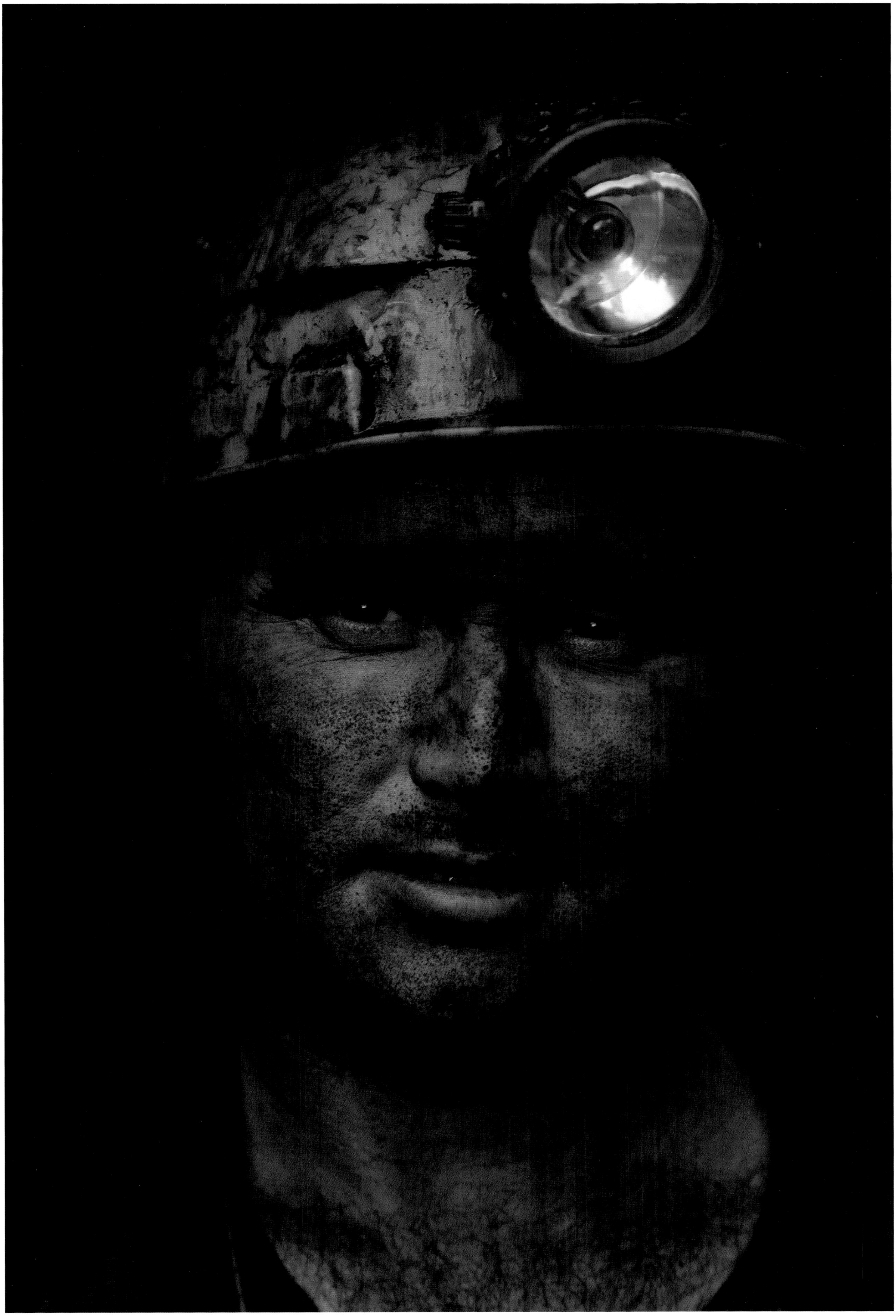

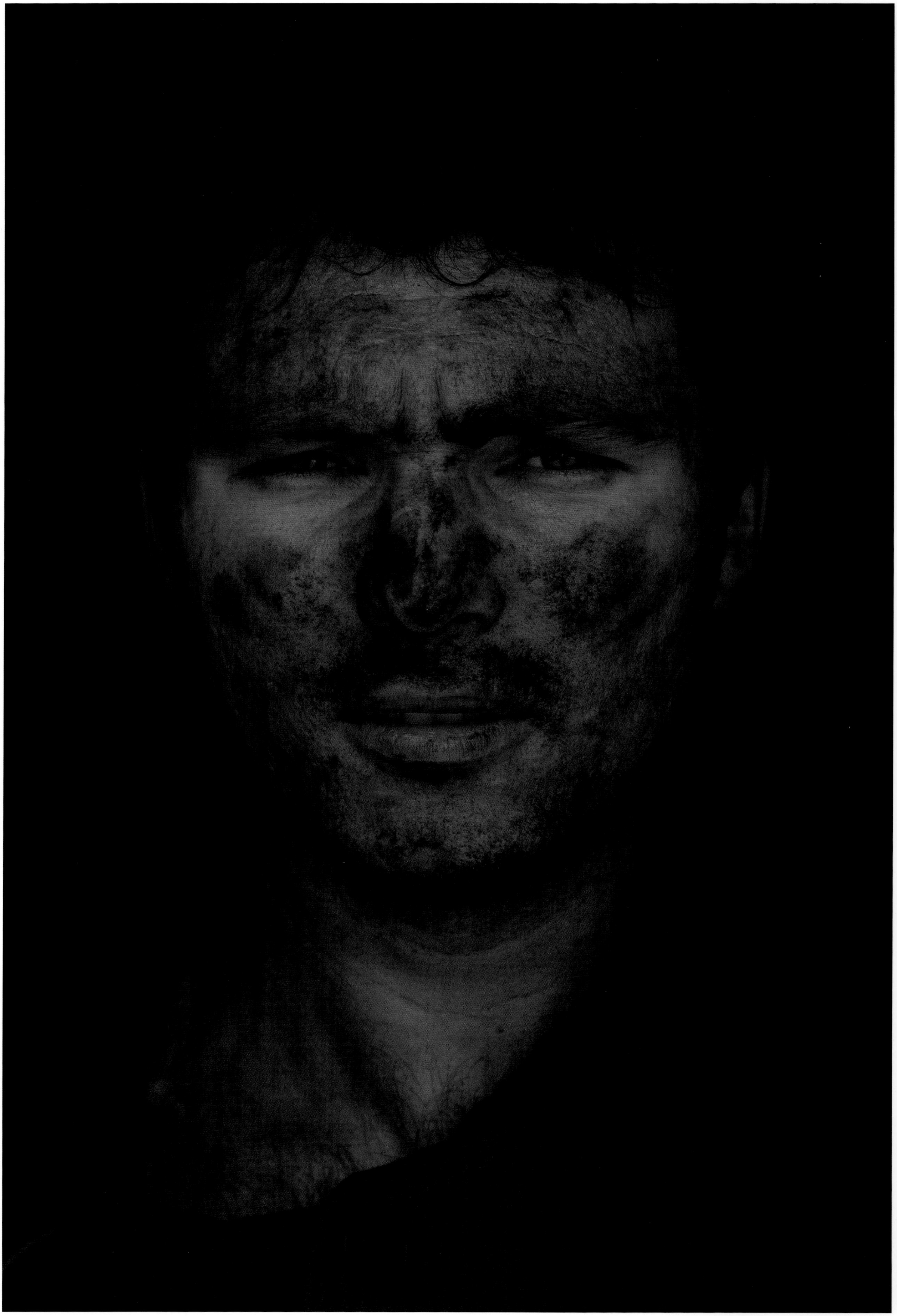

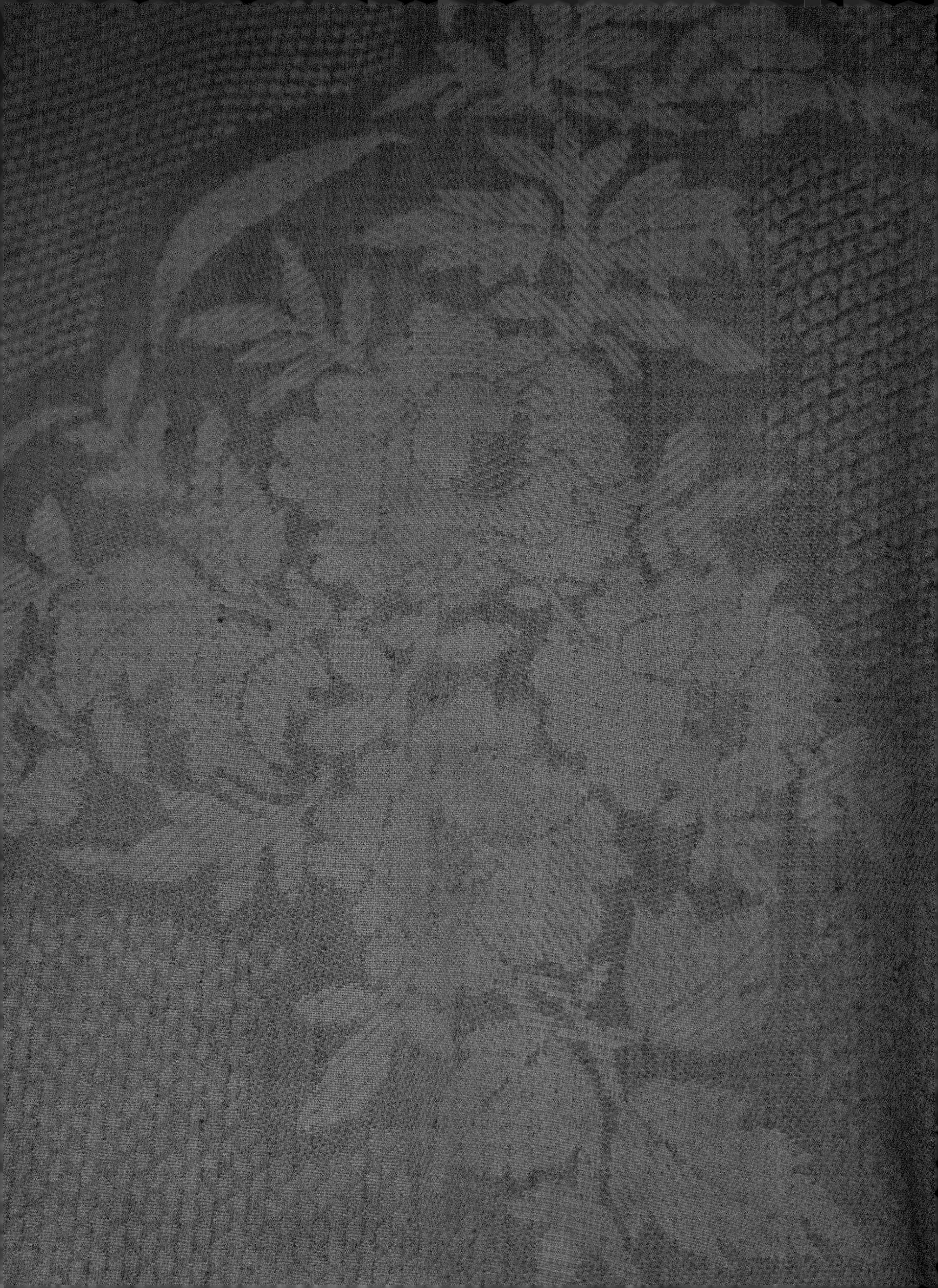

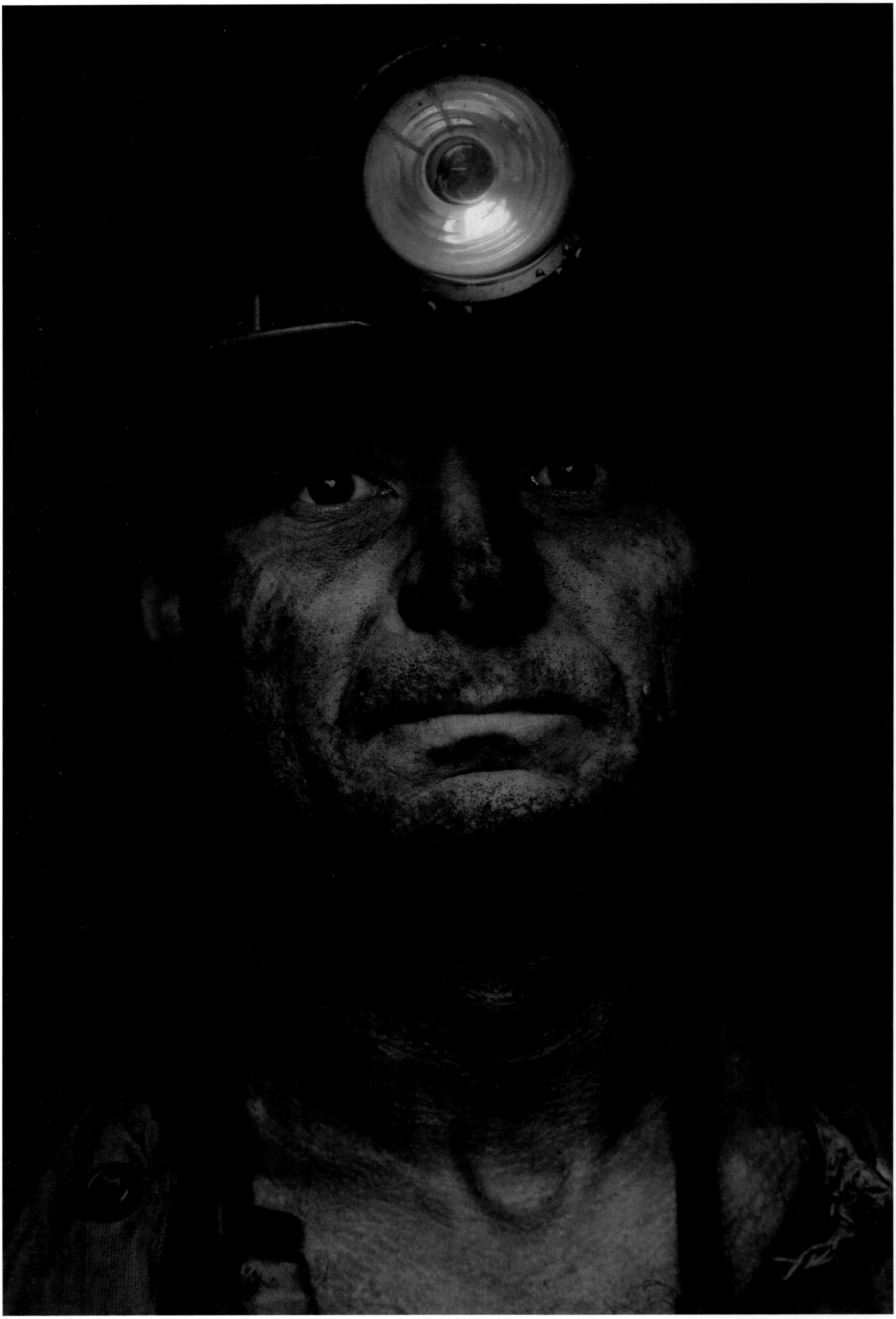

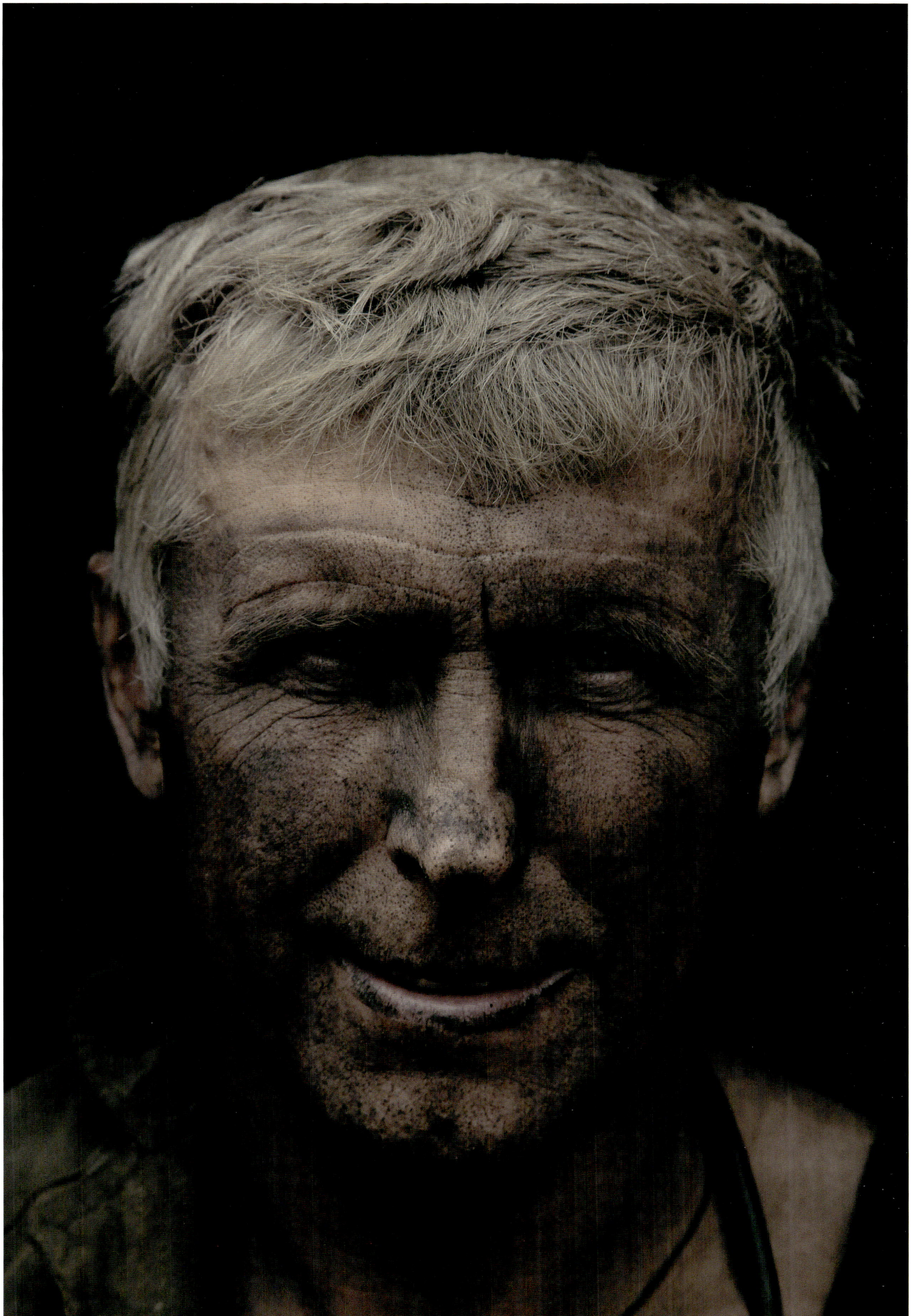

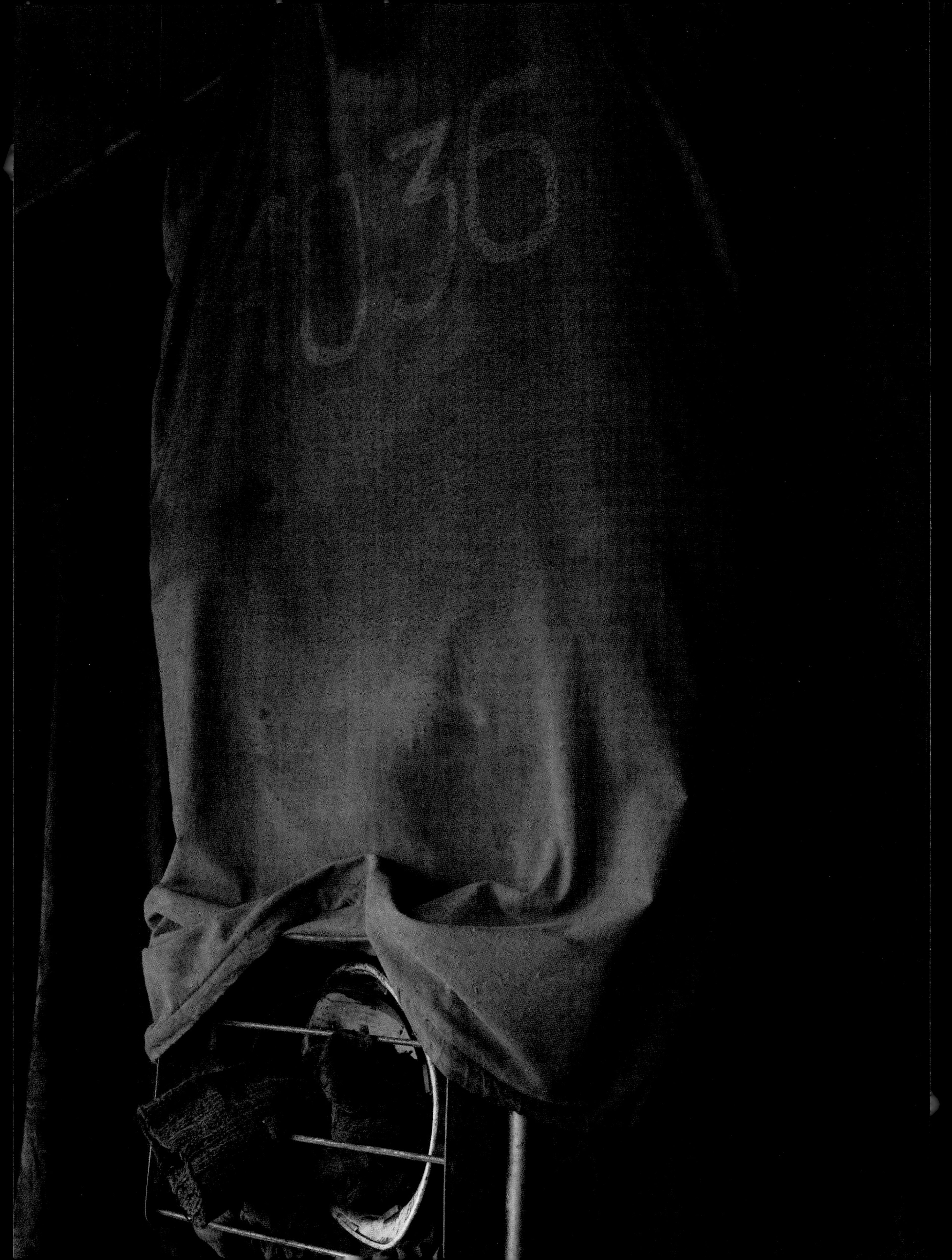
030

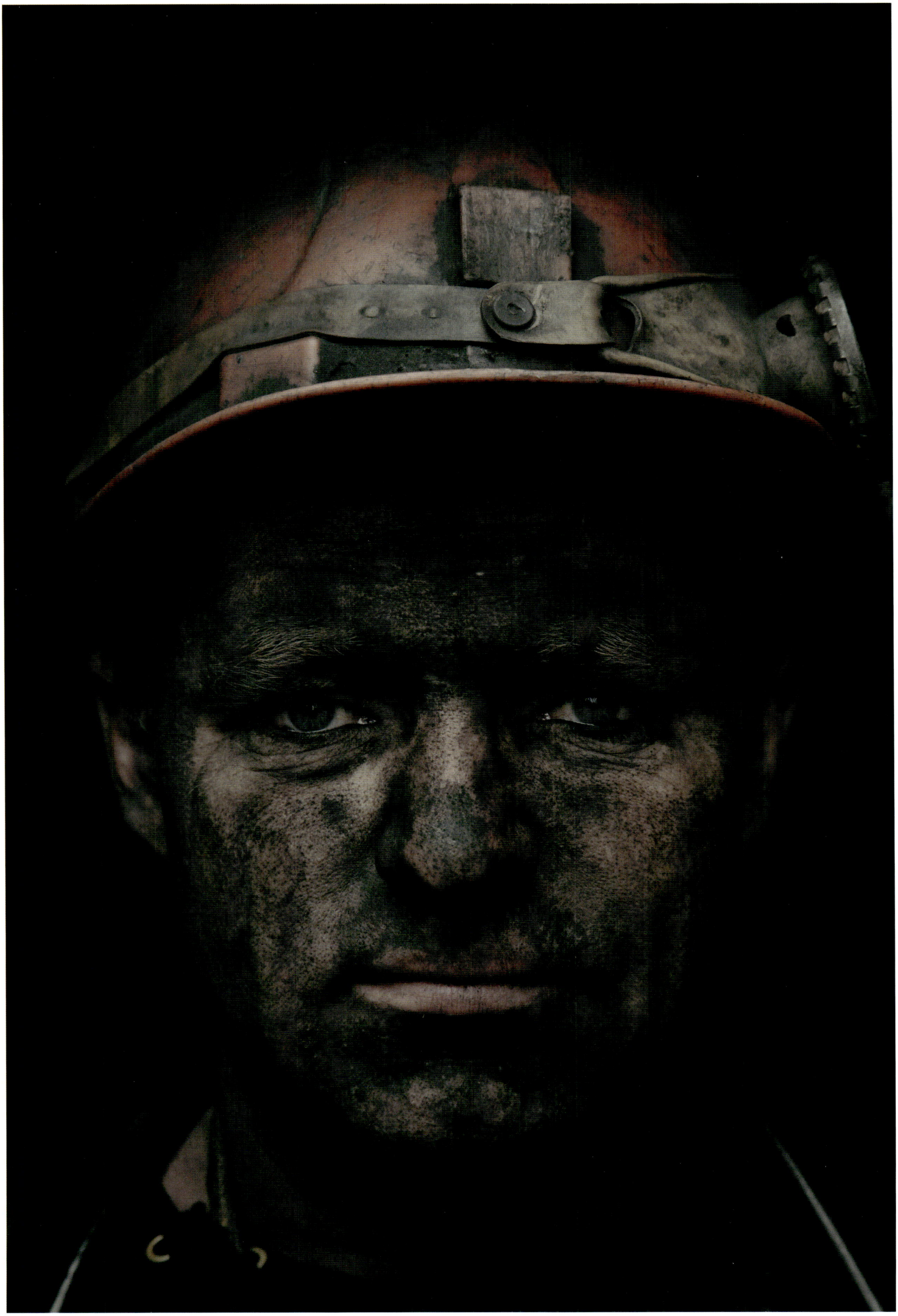

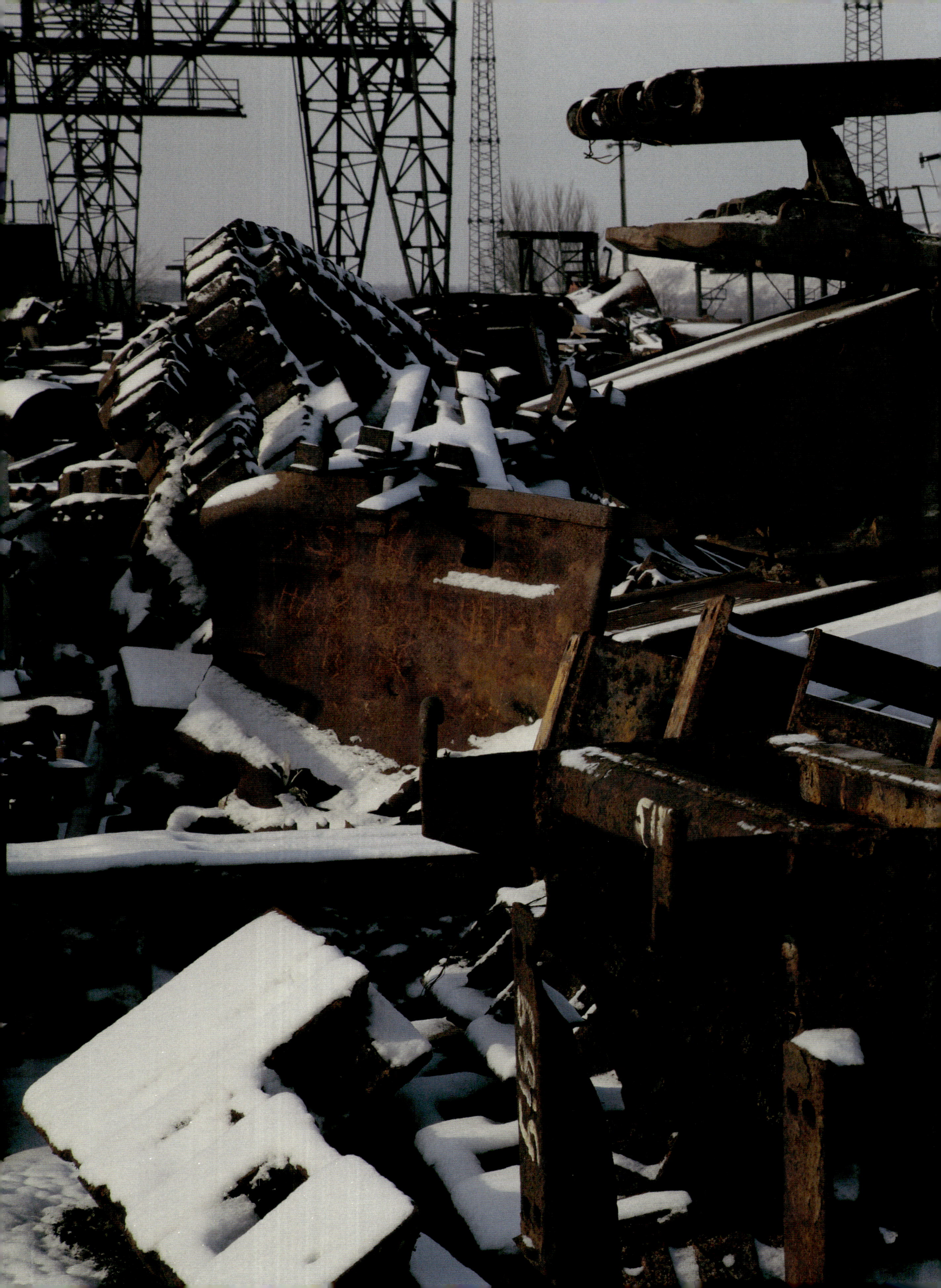

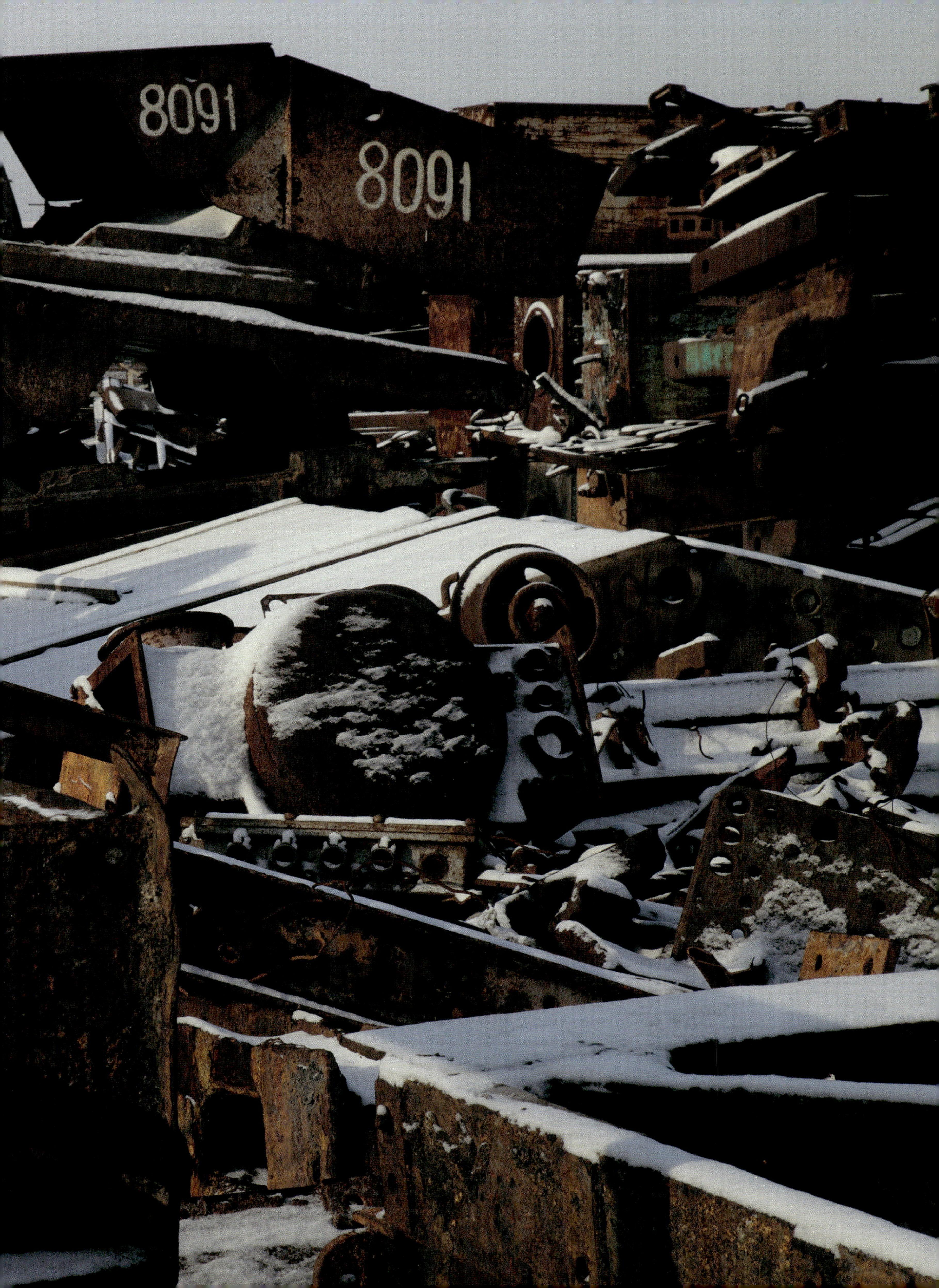
8091
8091

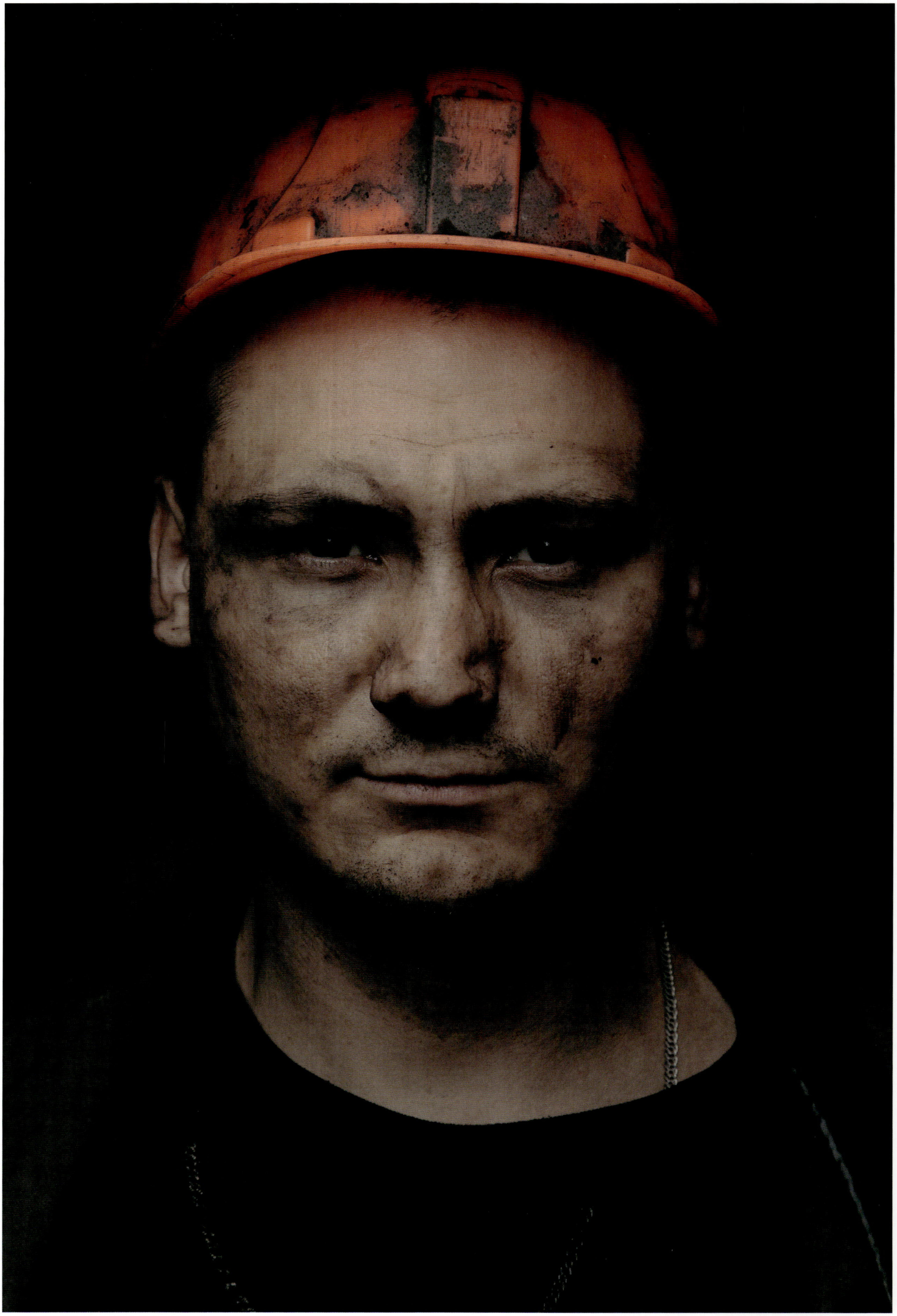

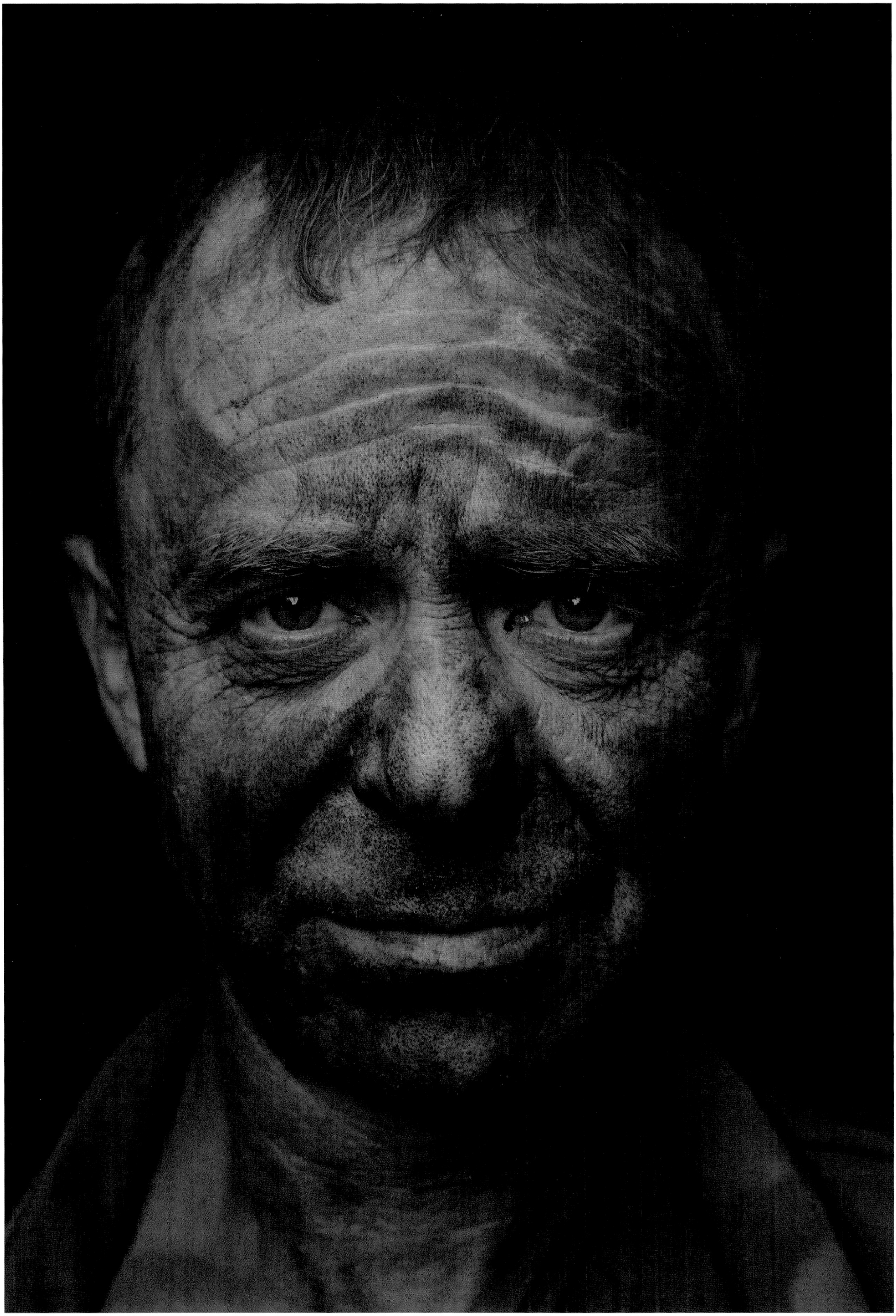

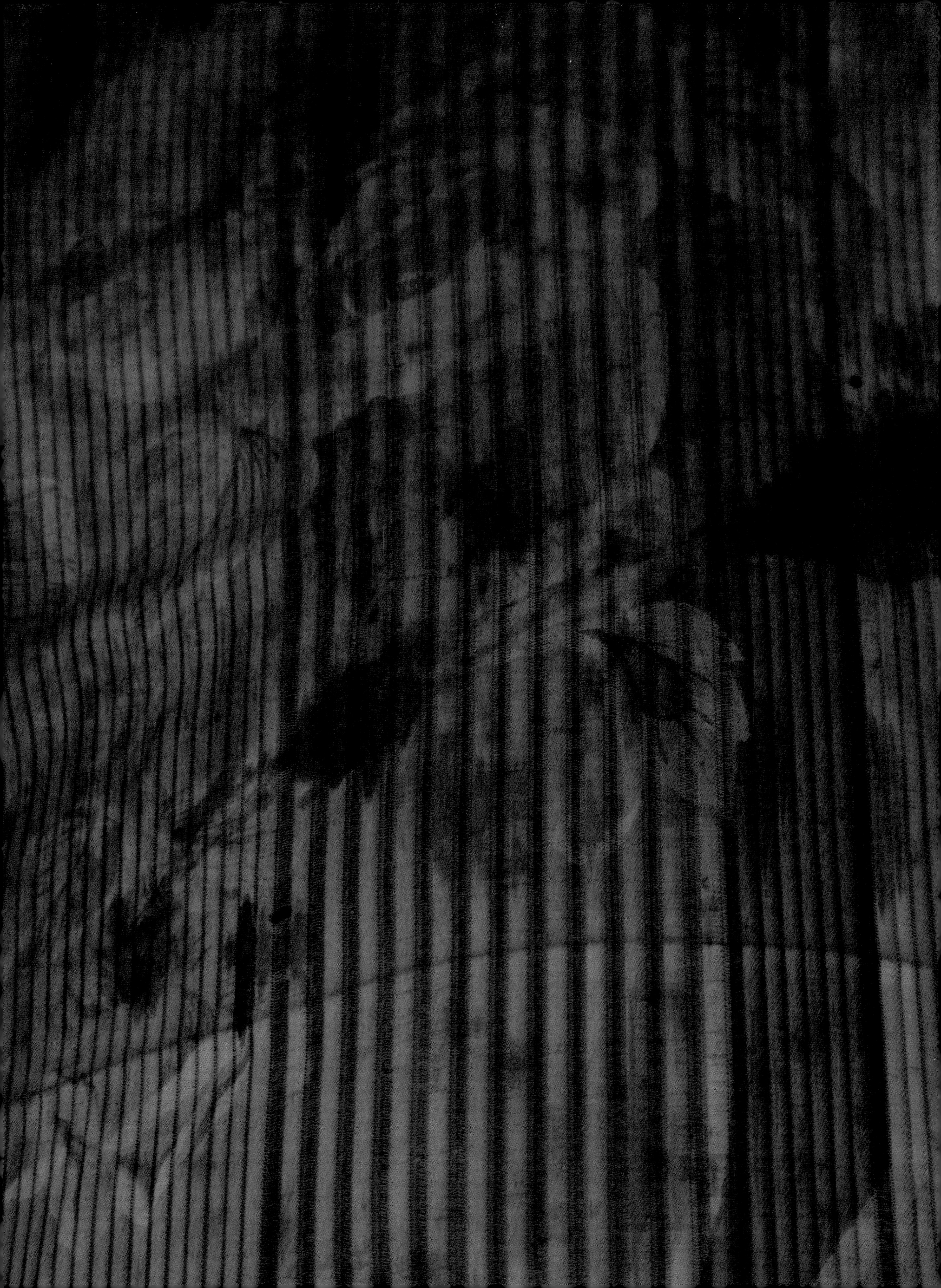

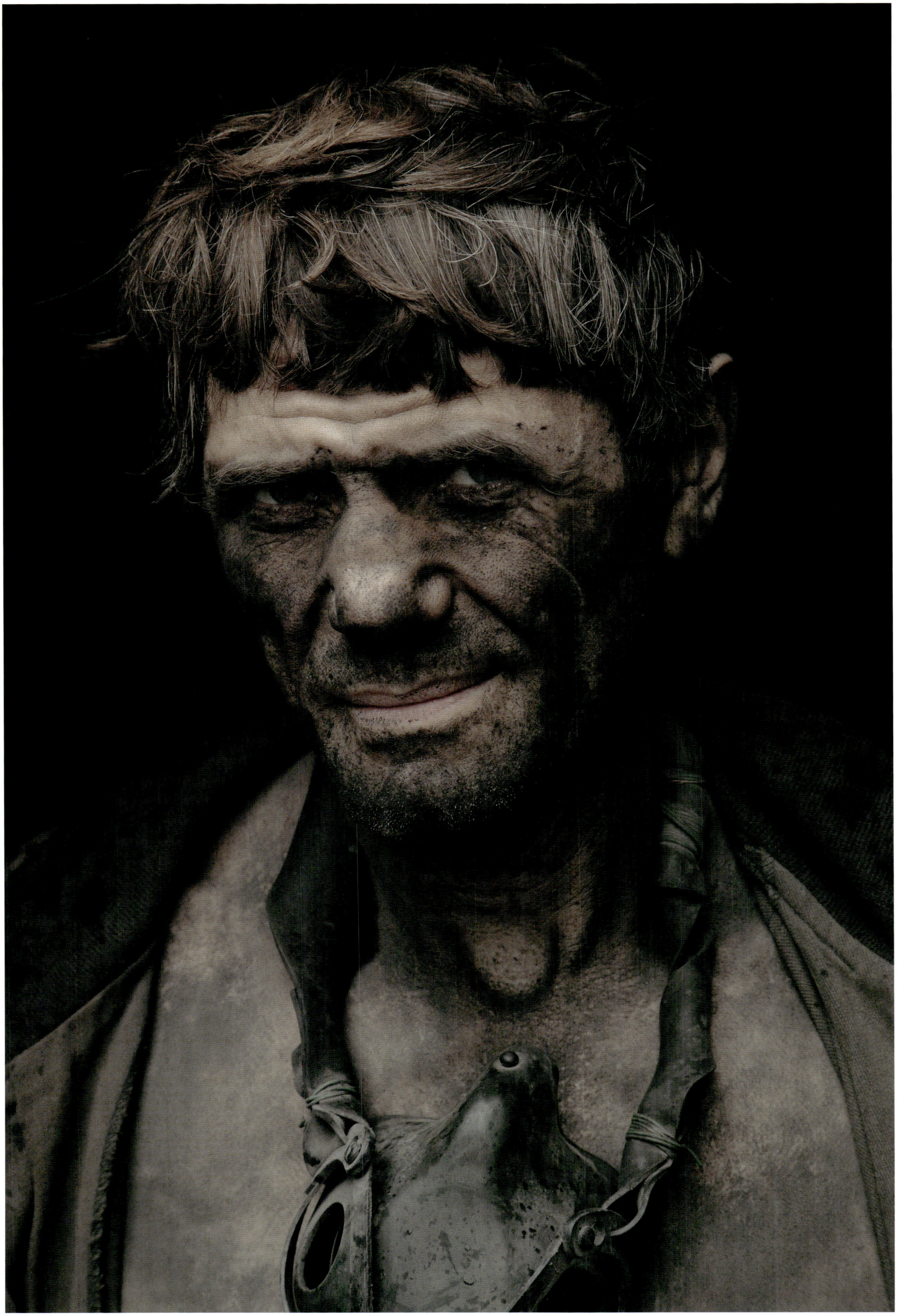

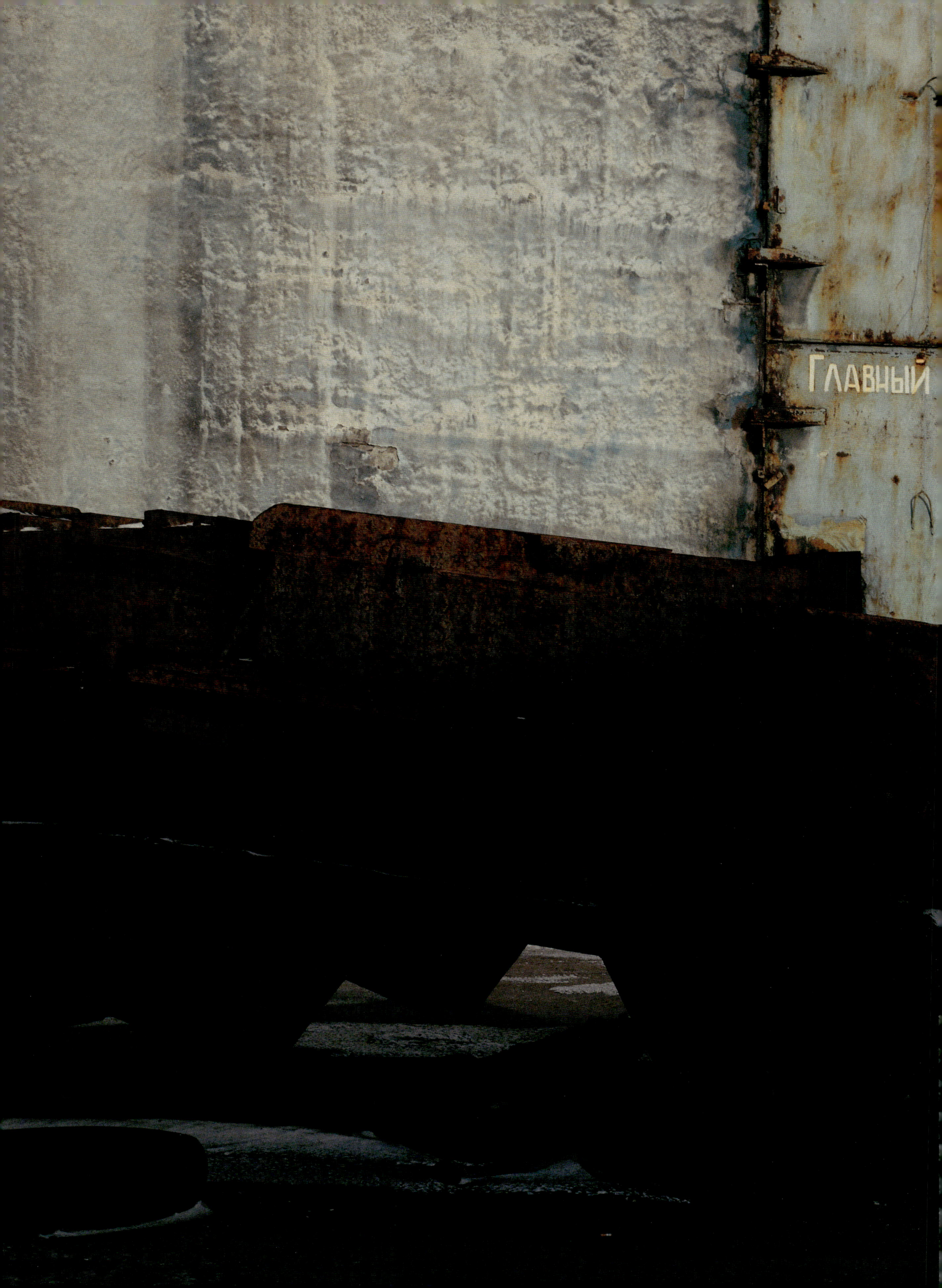
Главный

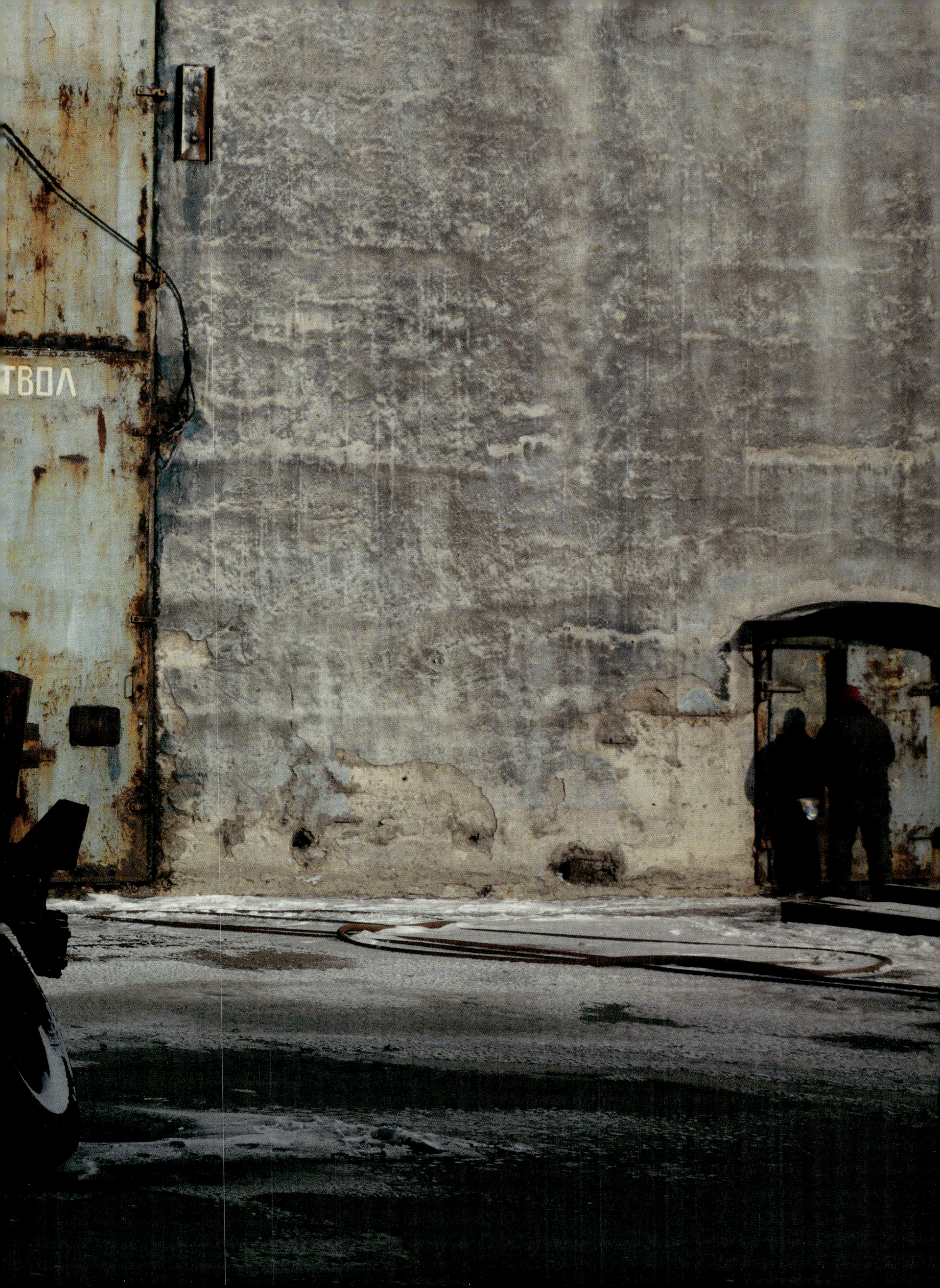
ТВОЛ

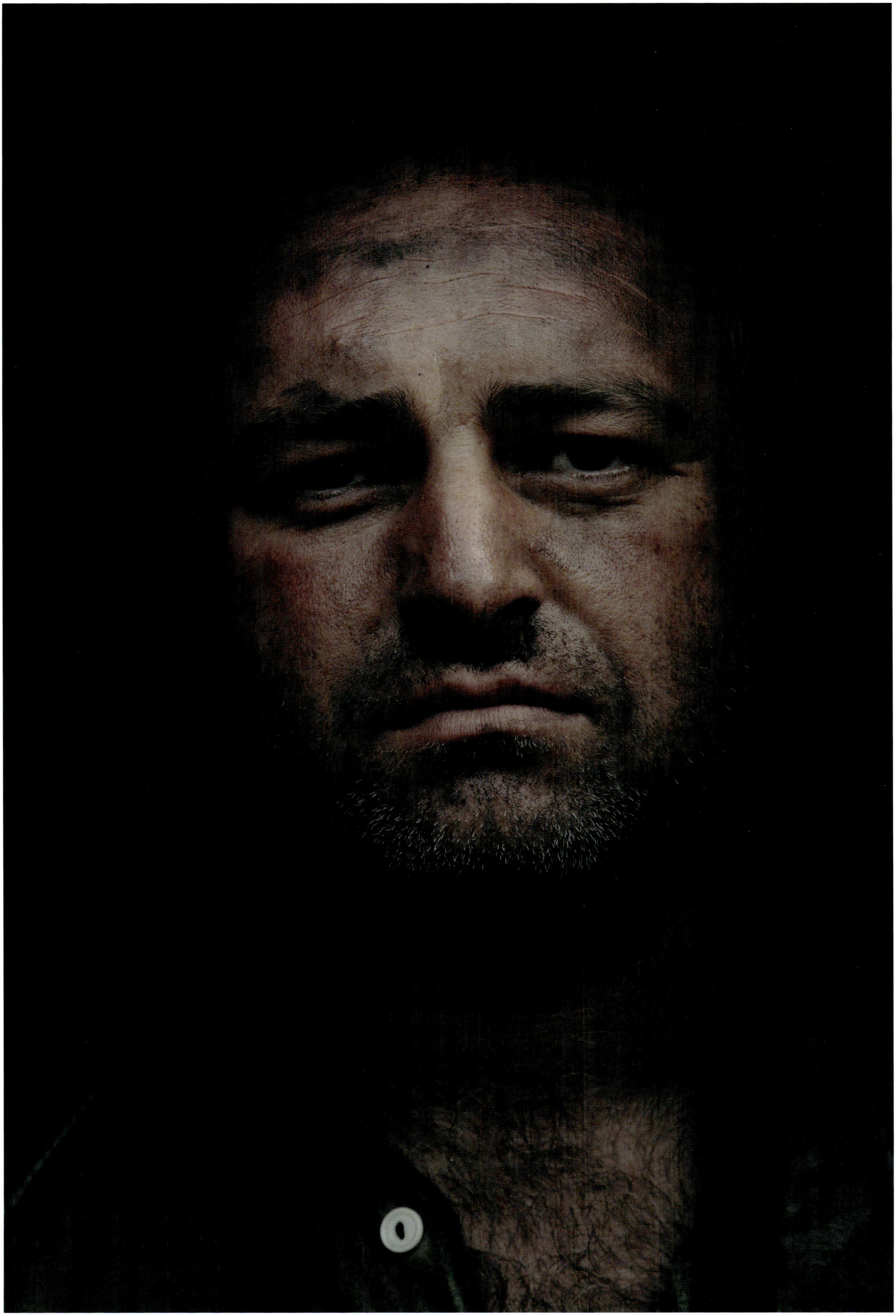

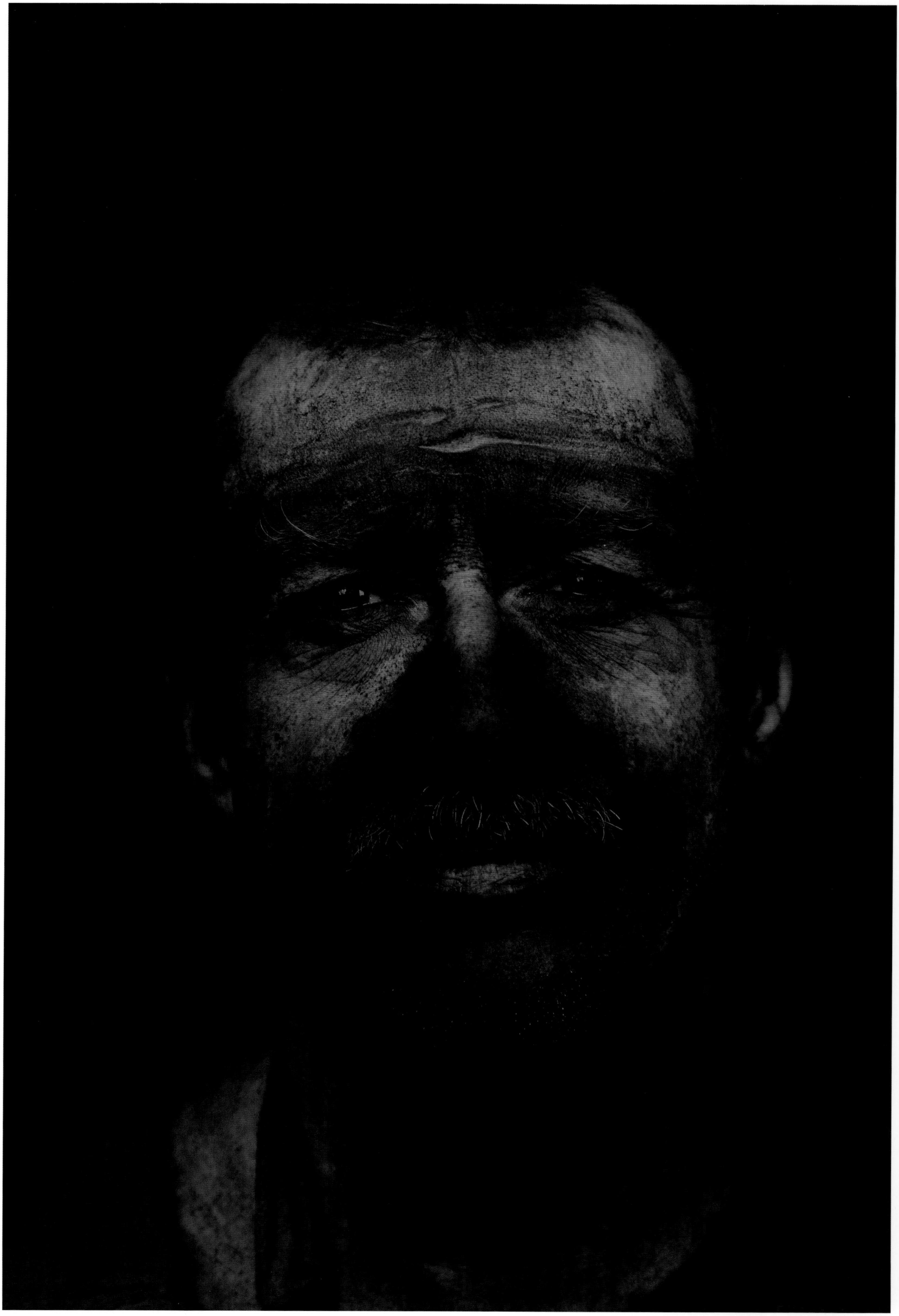

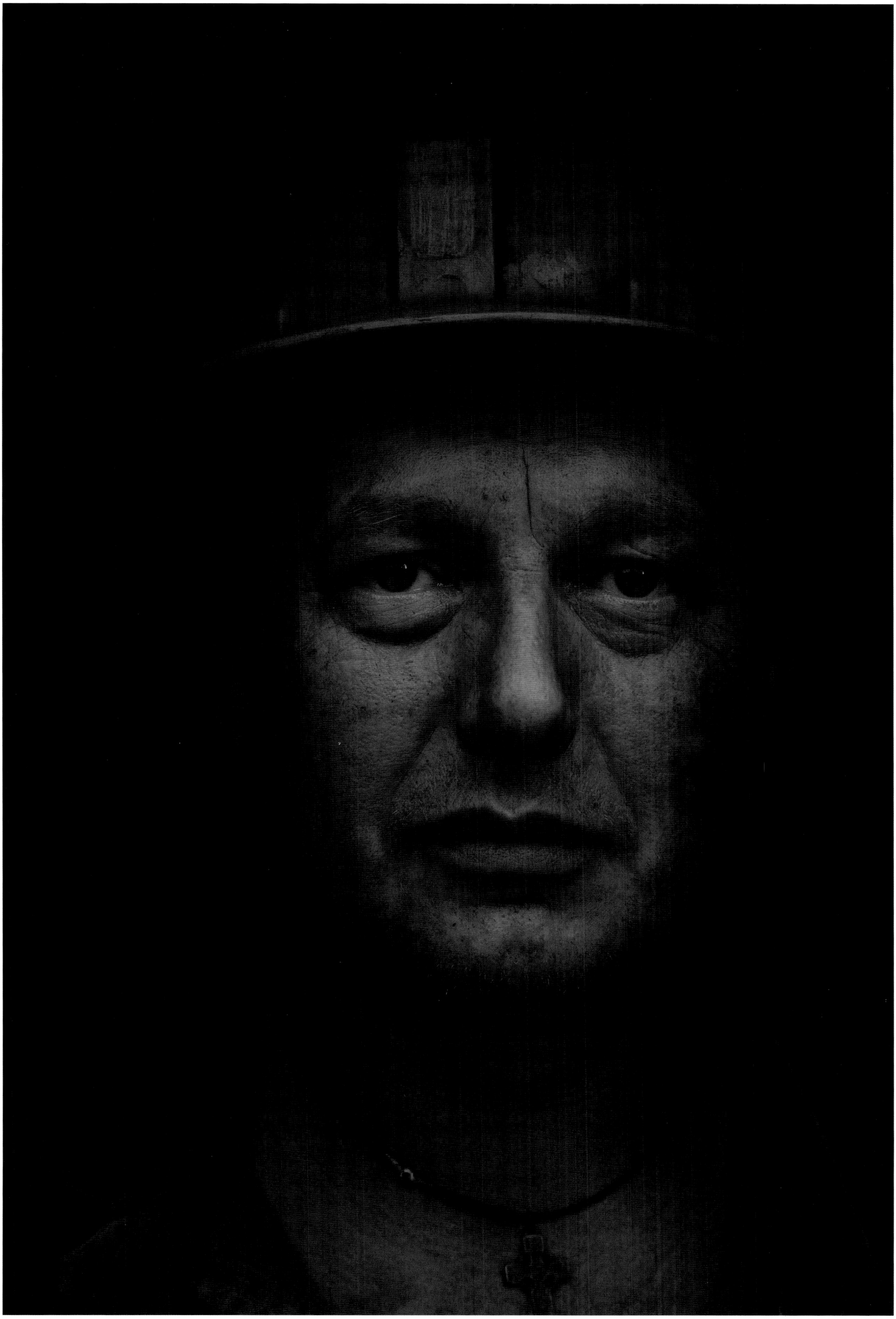

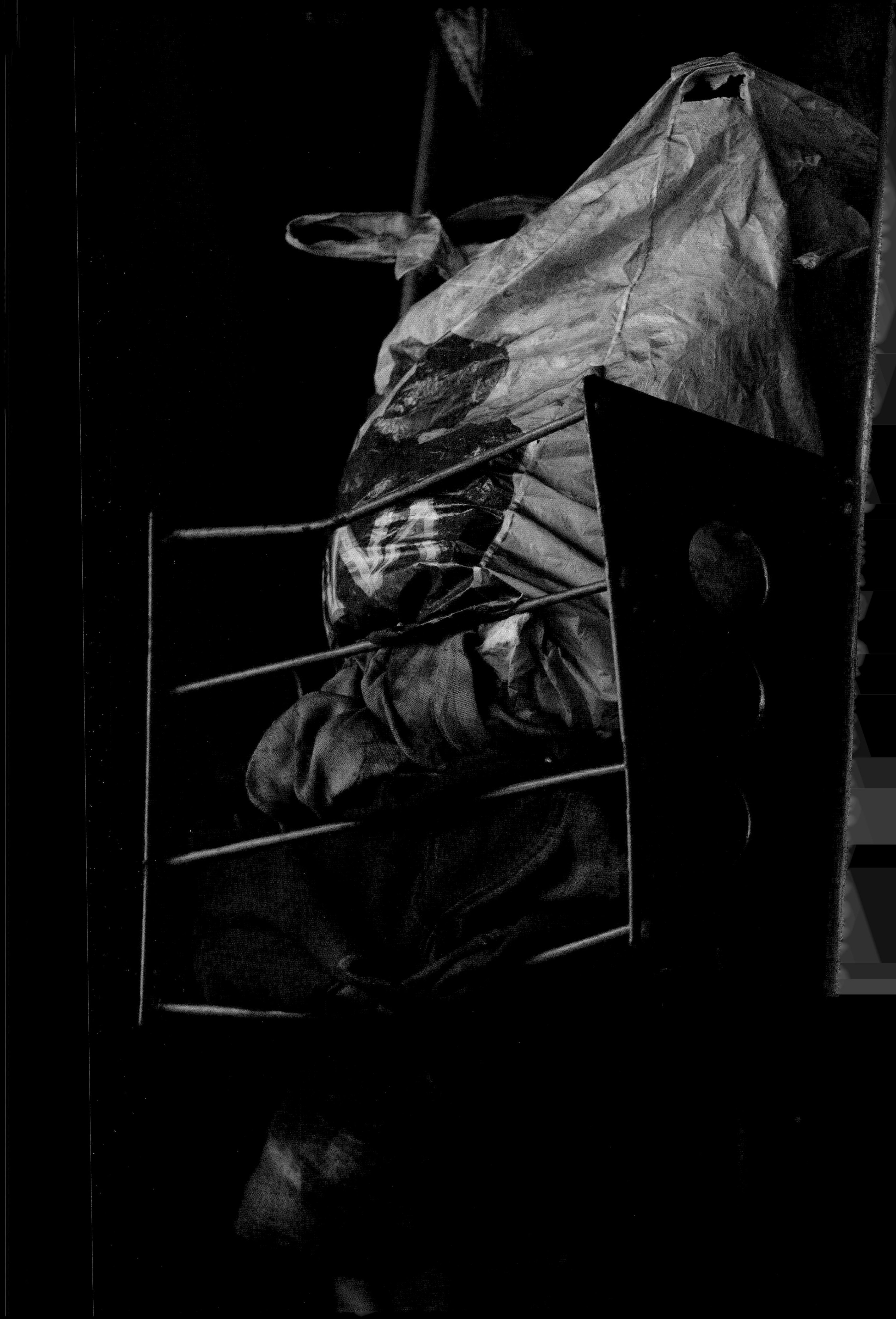

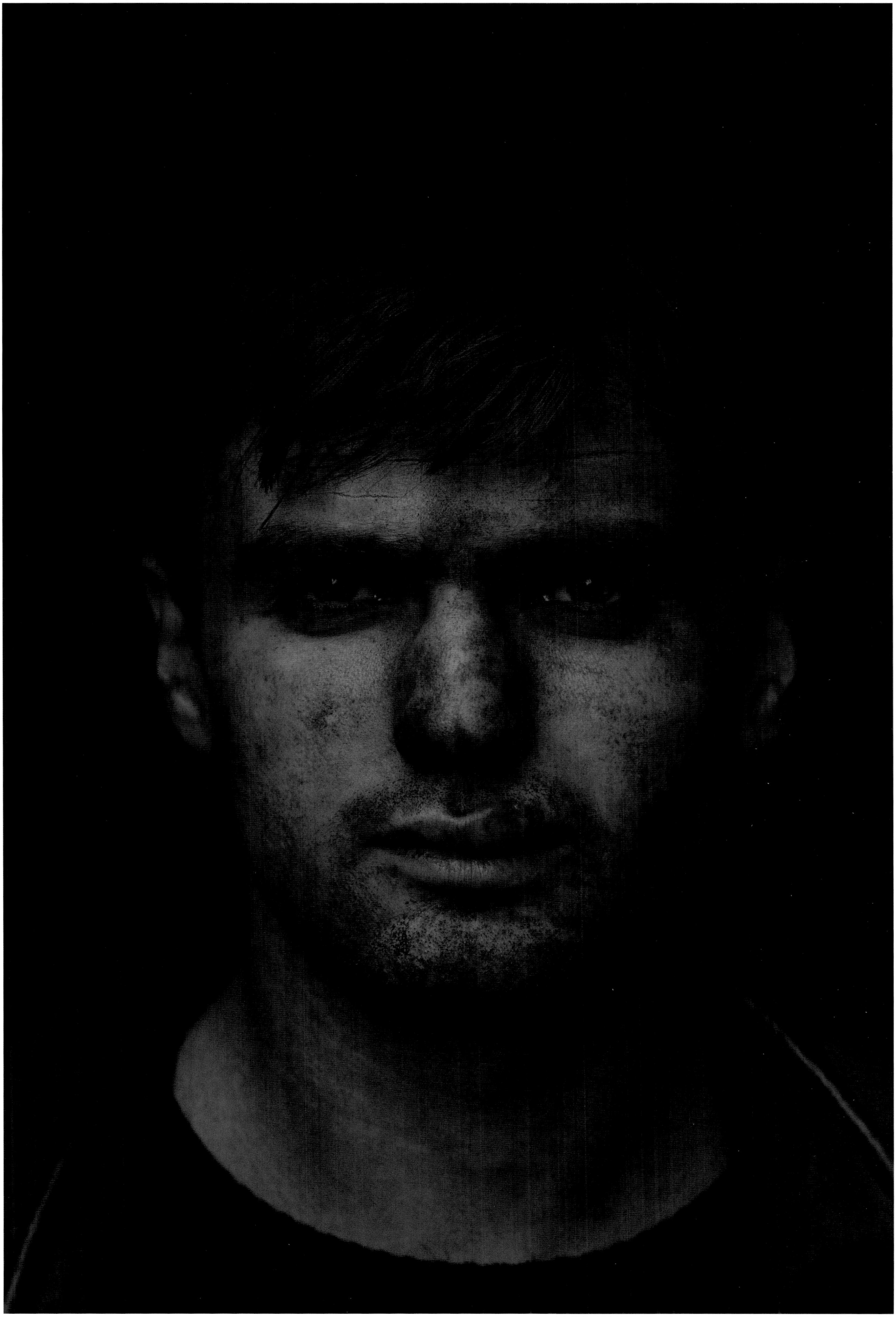

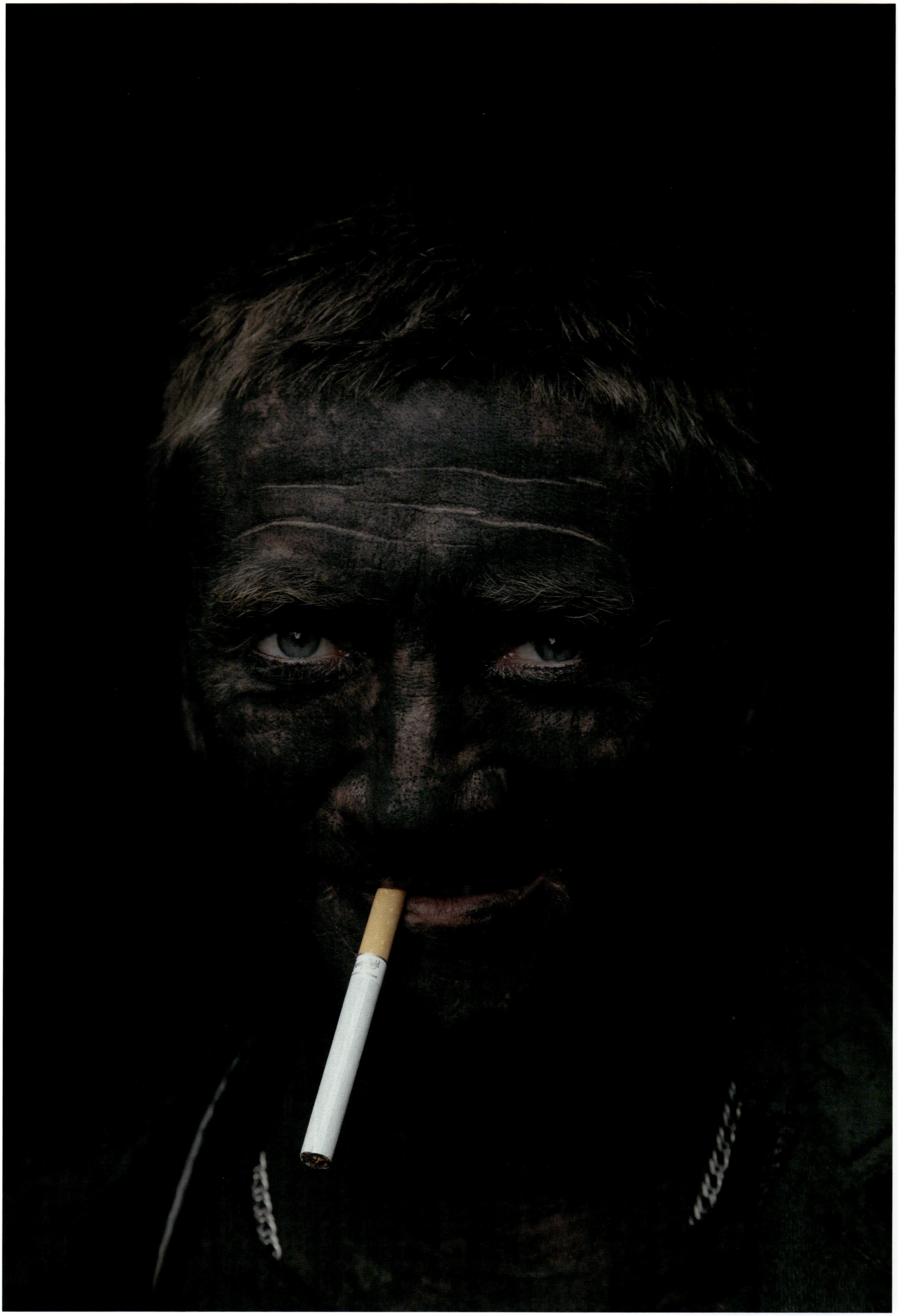

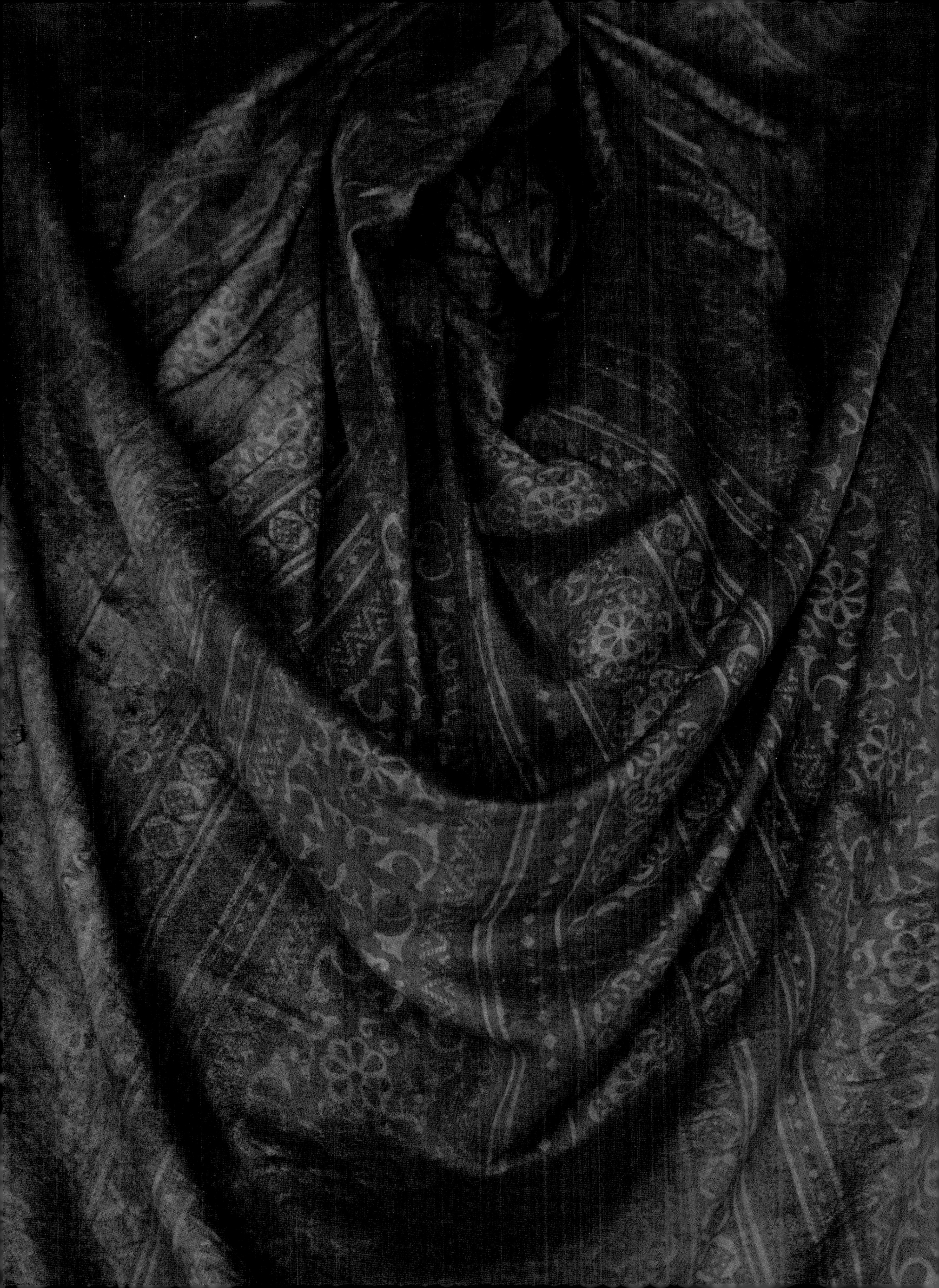

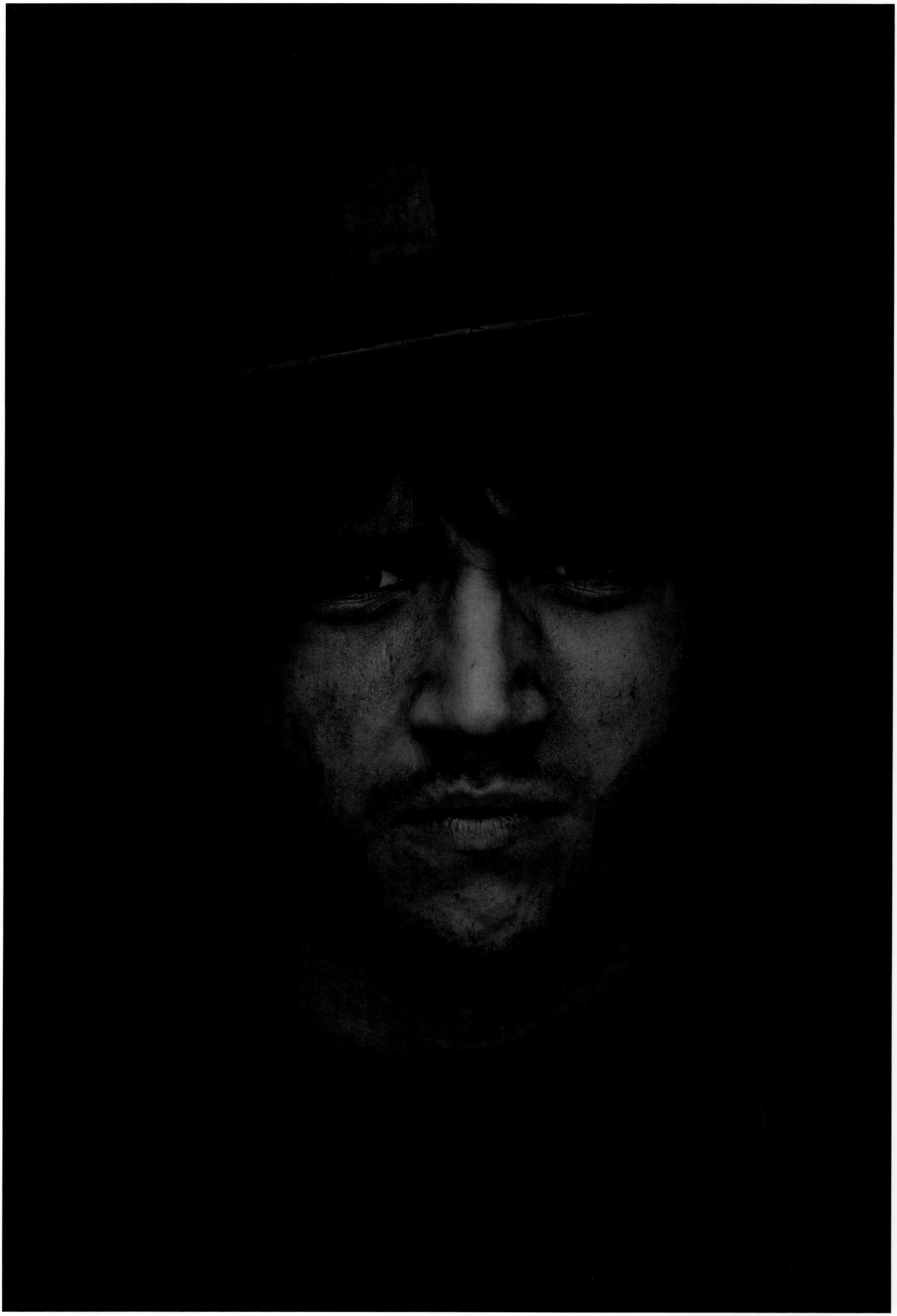

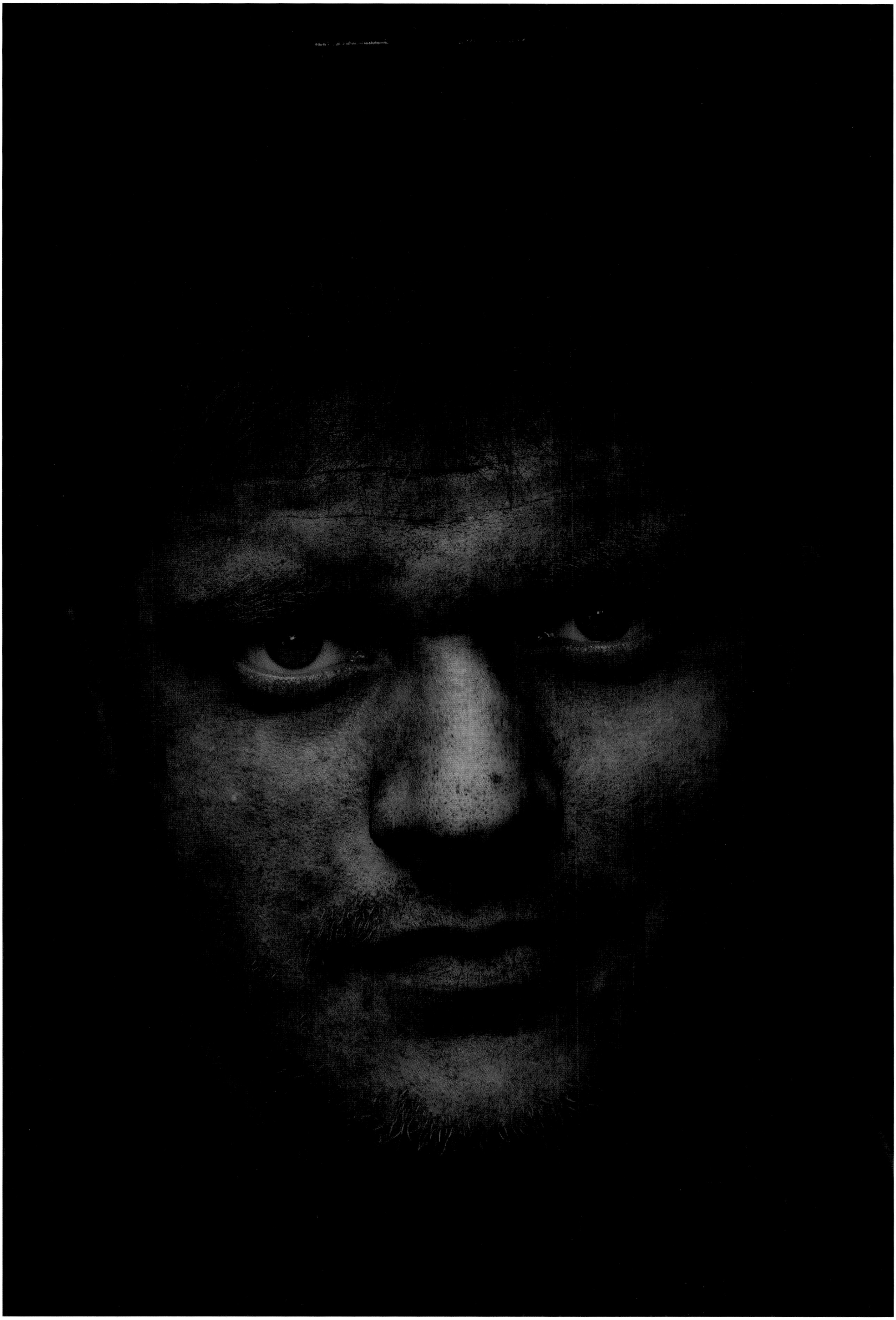

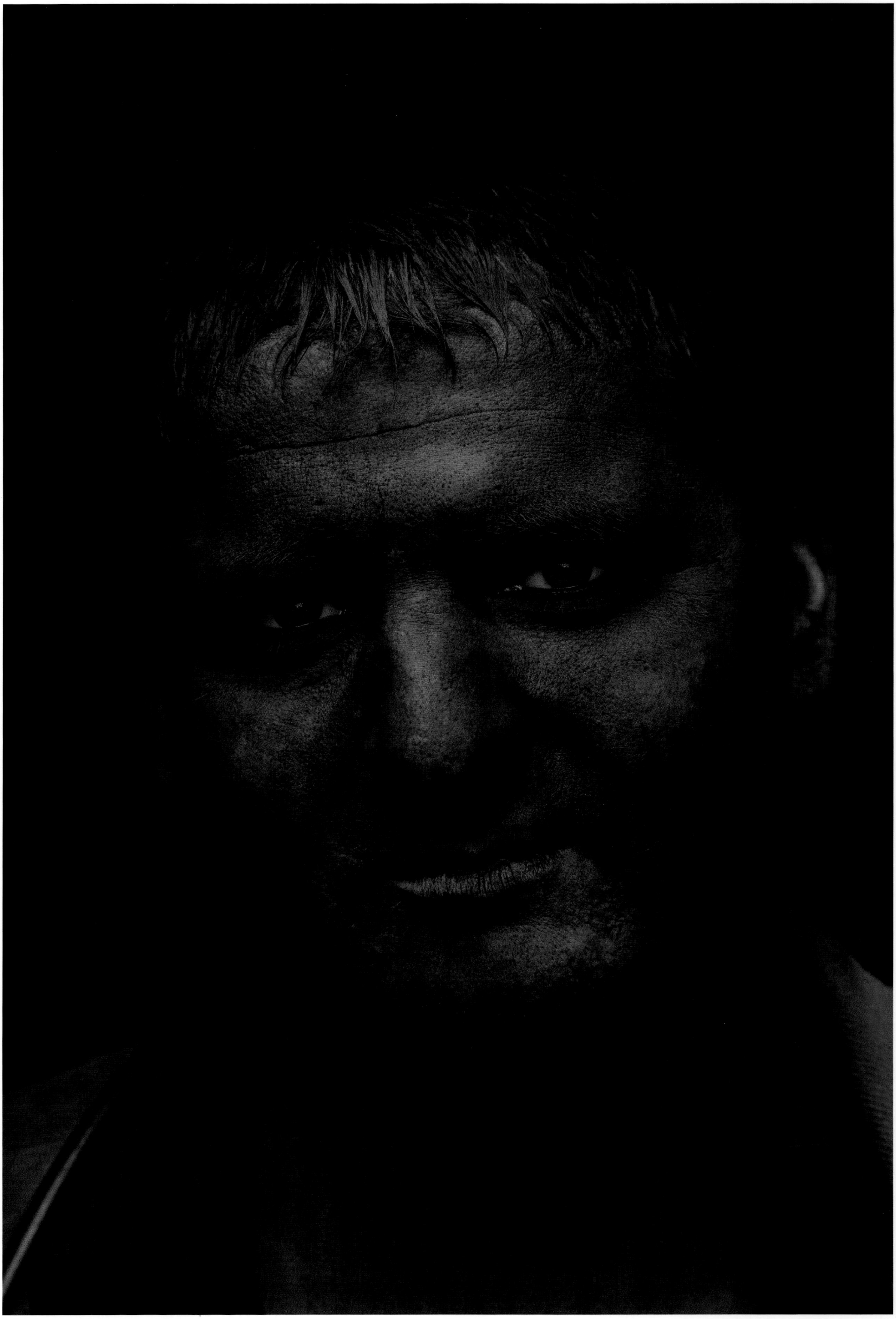

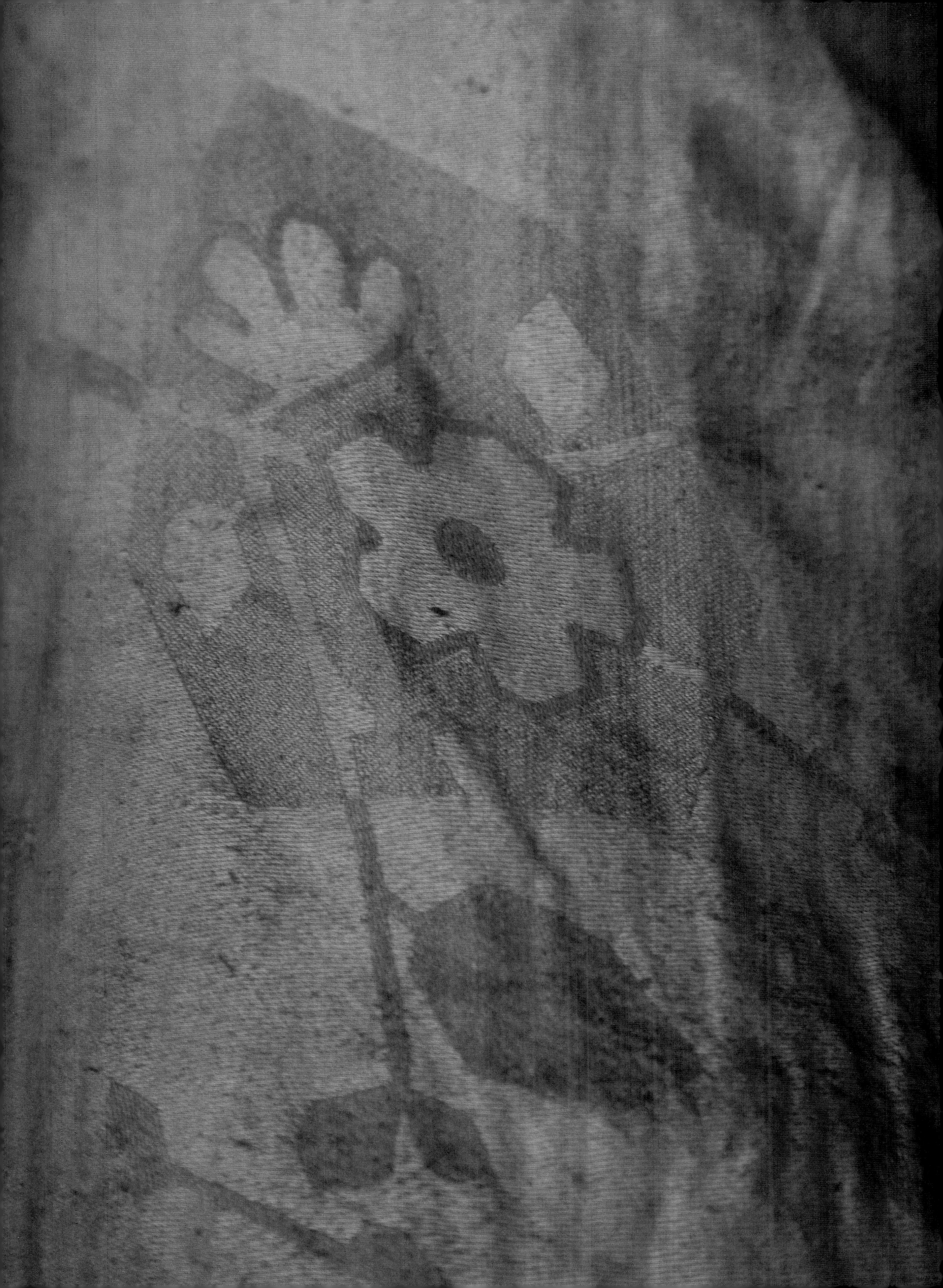

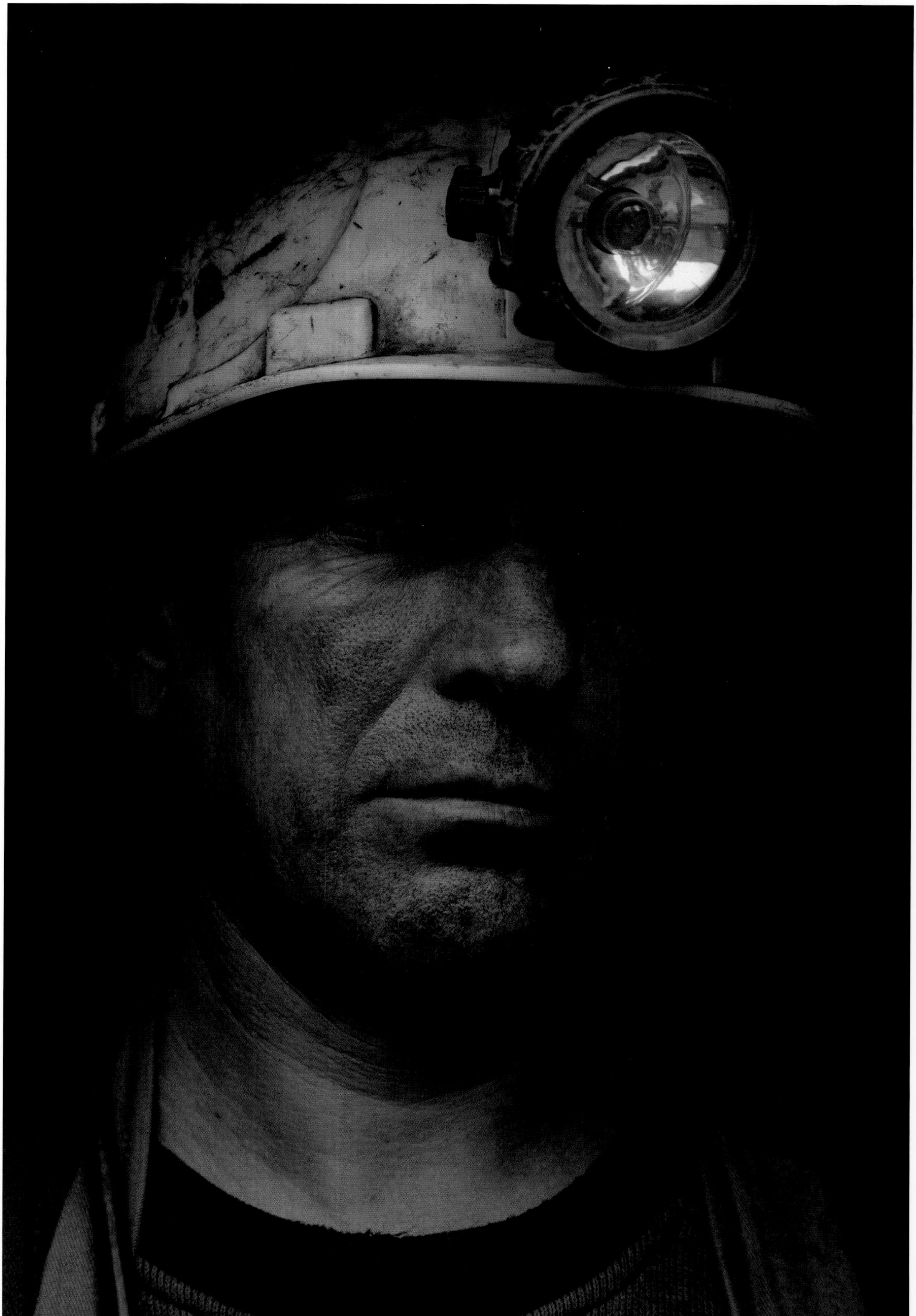

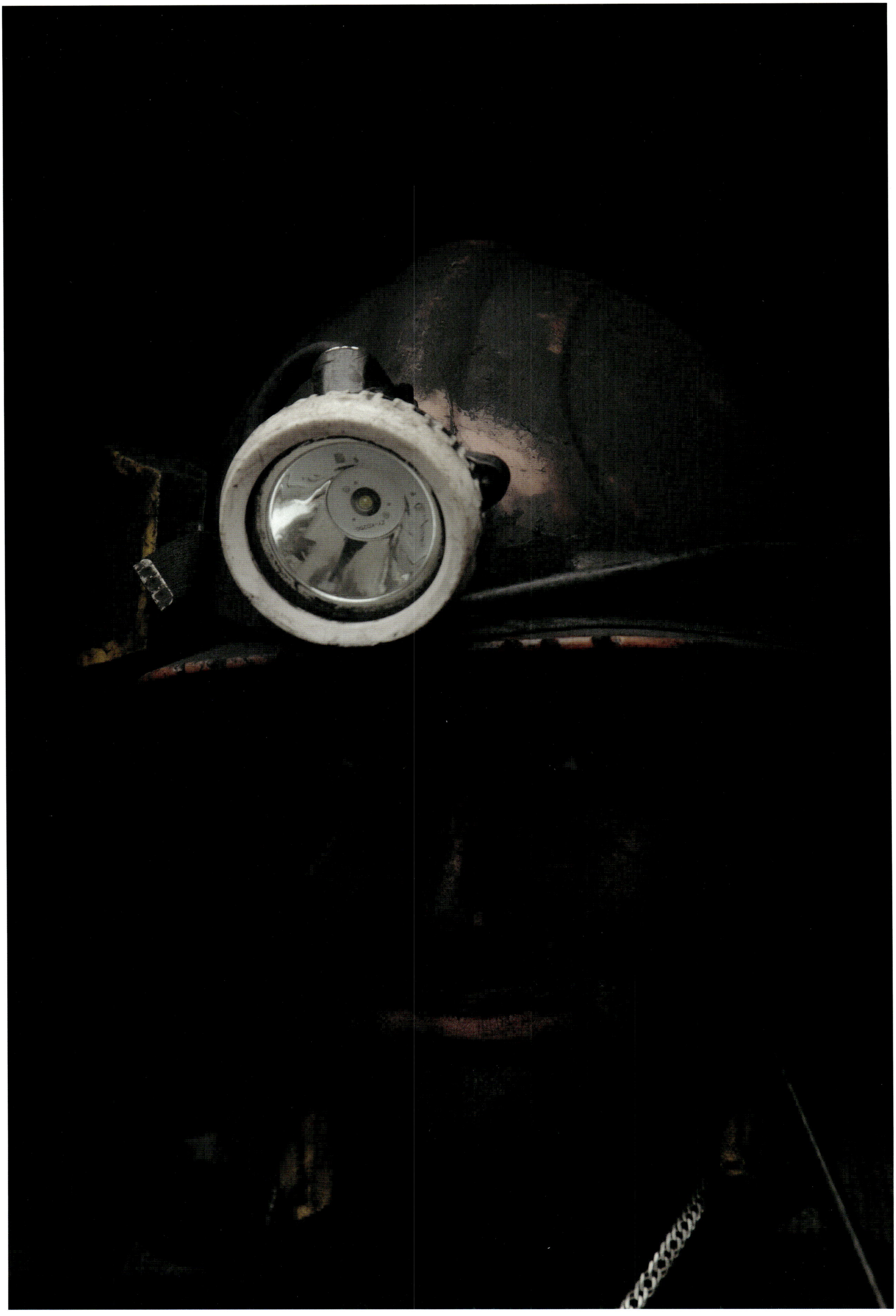

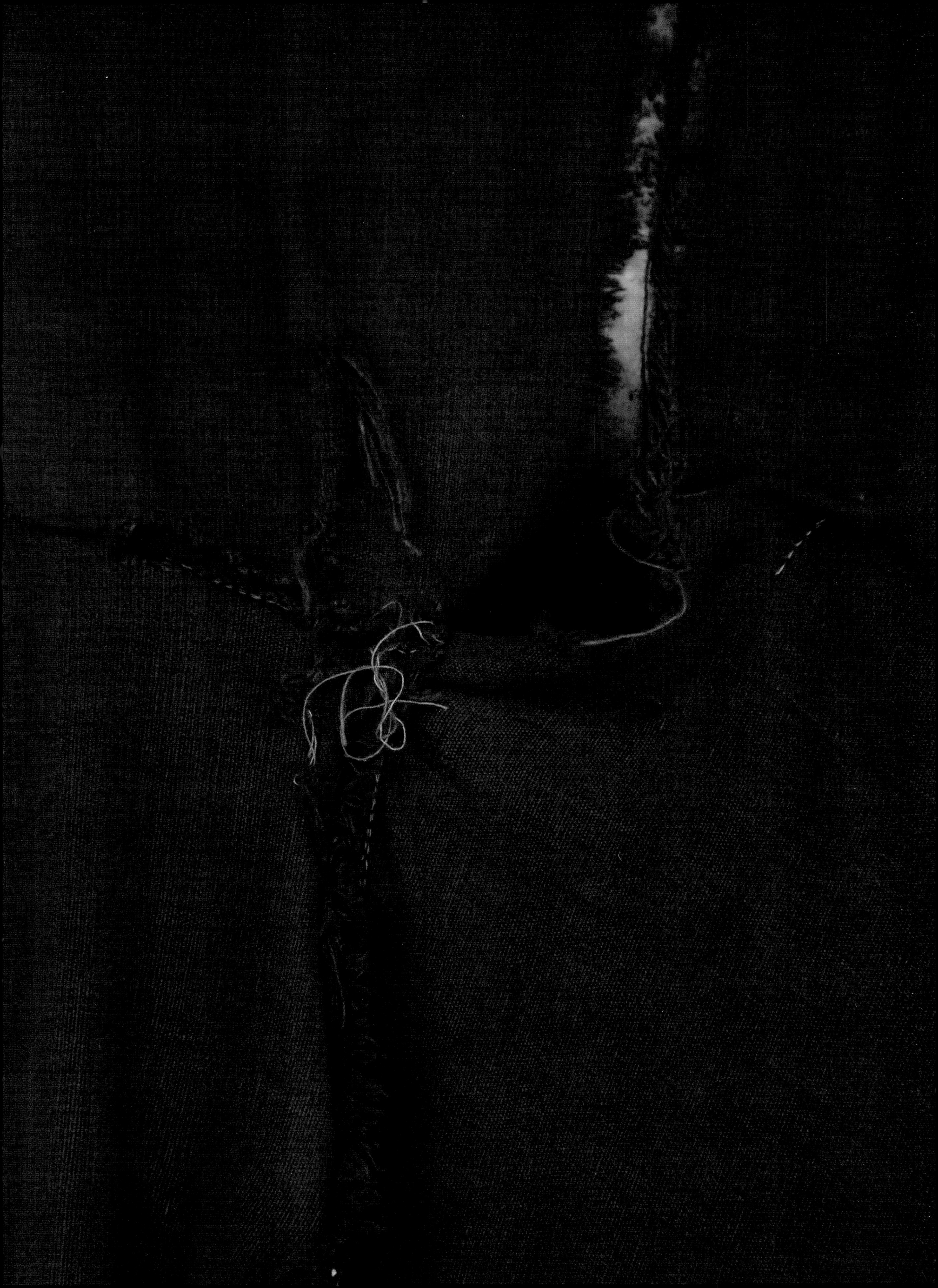

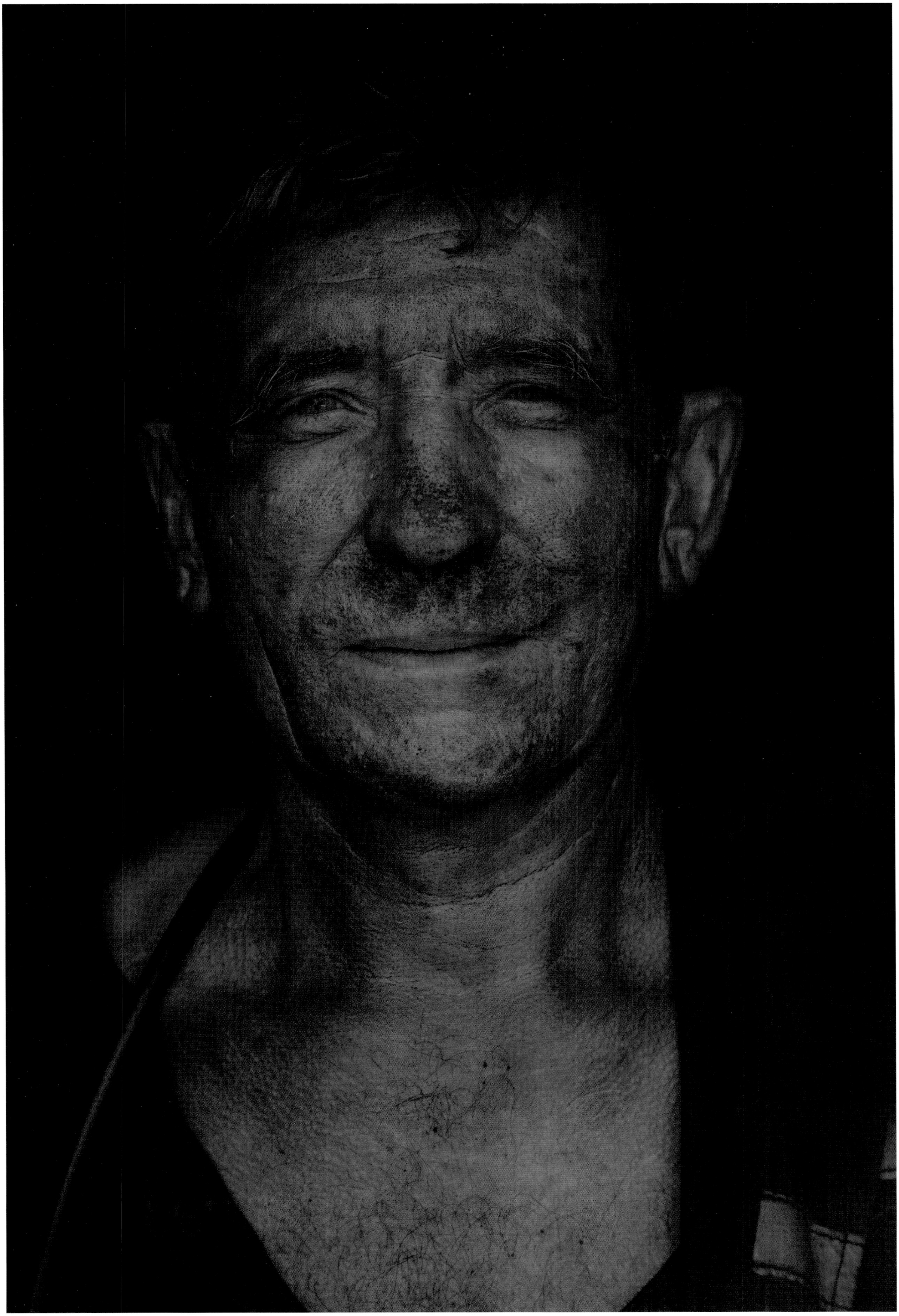

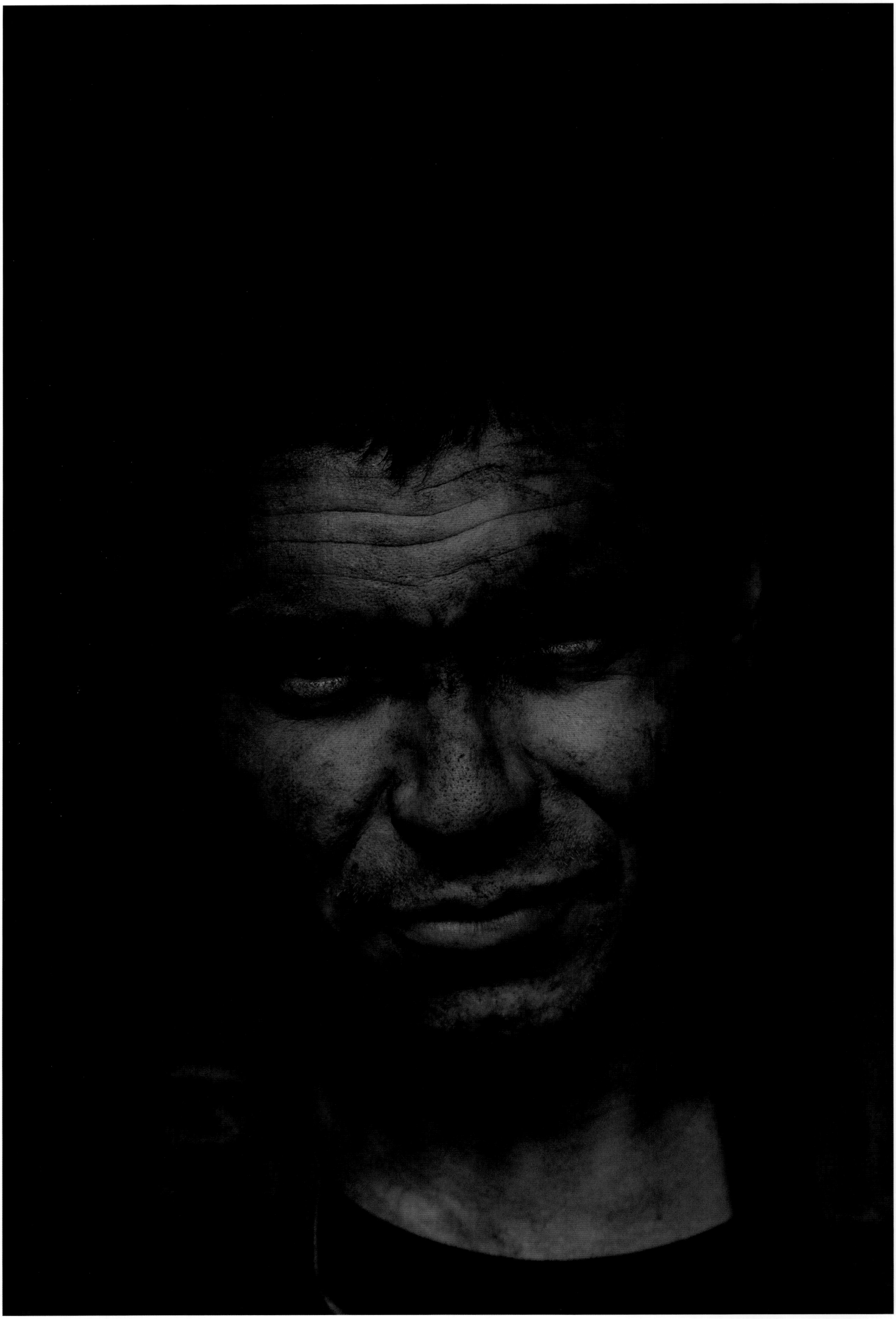

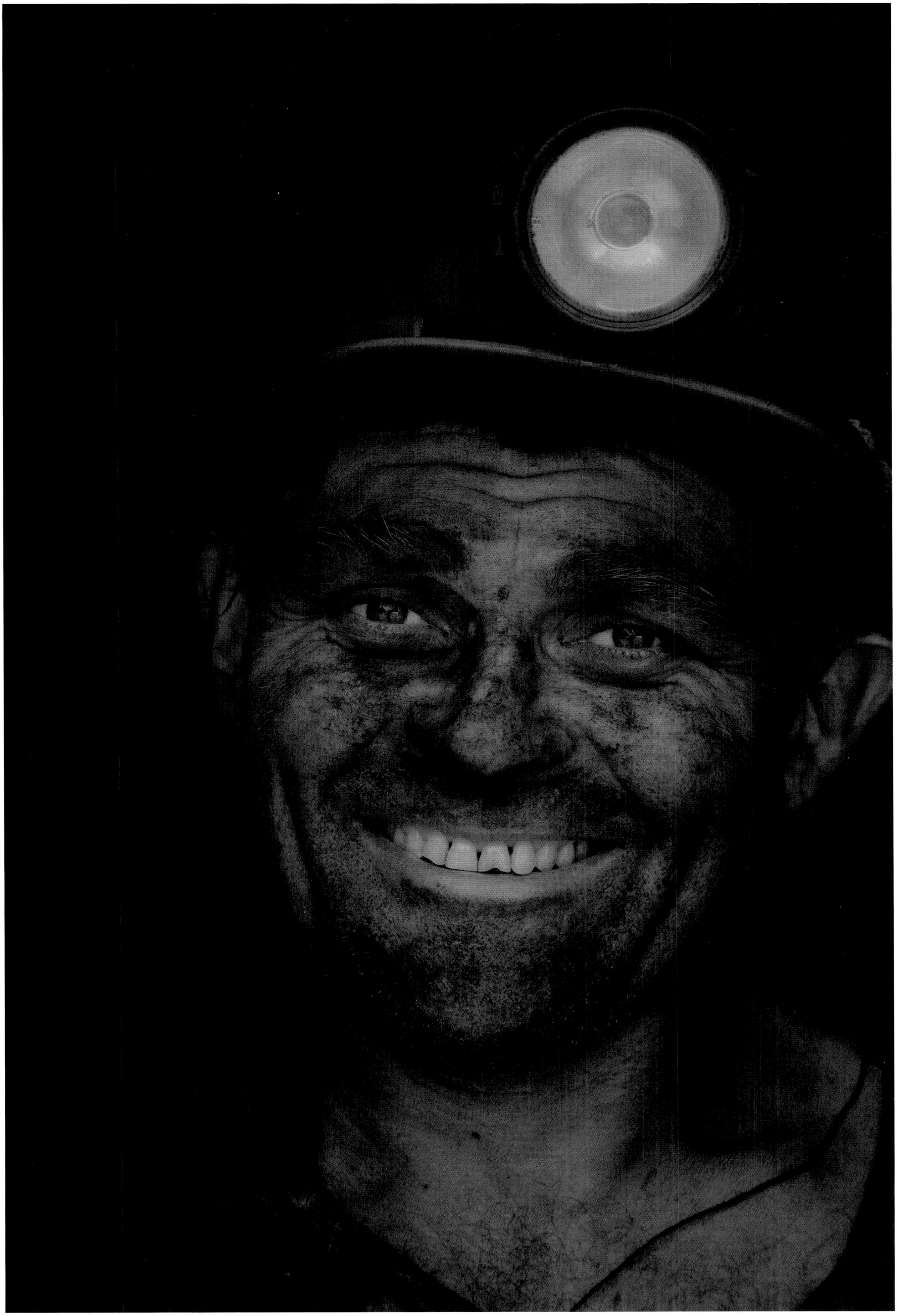

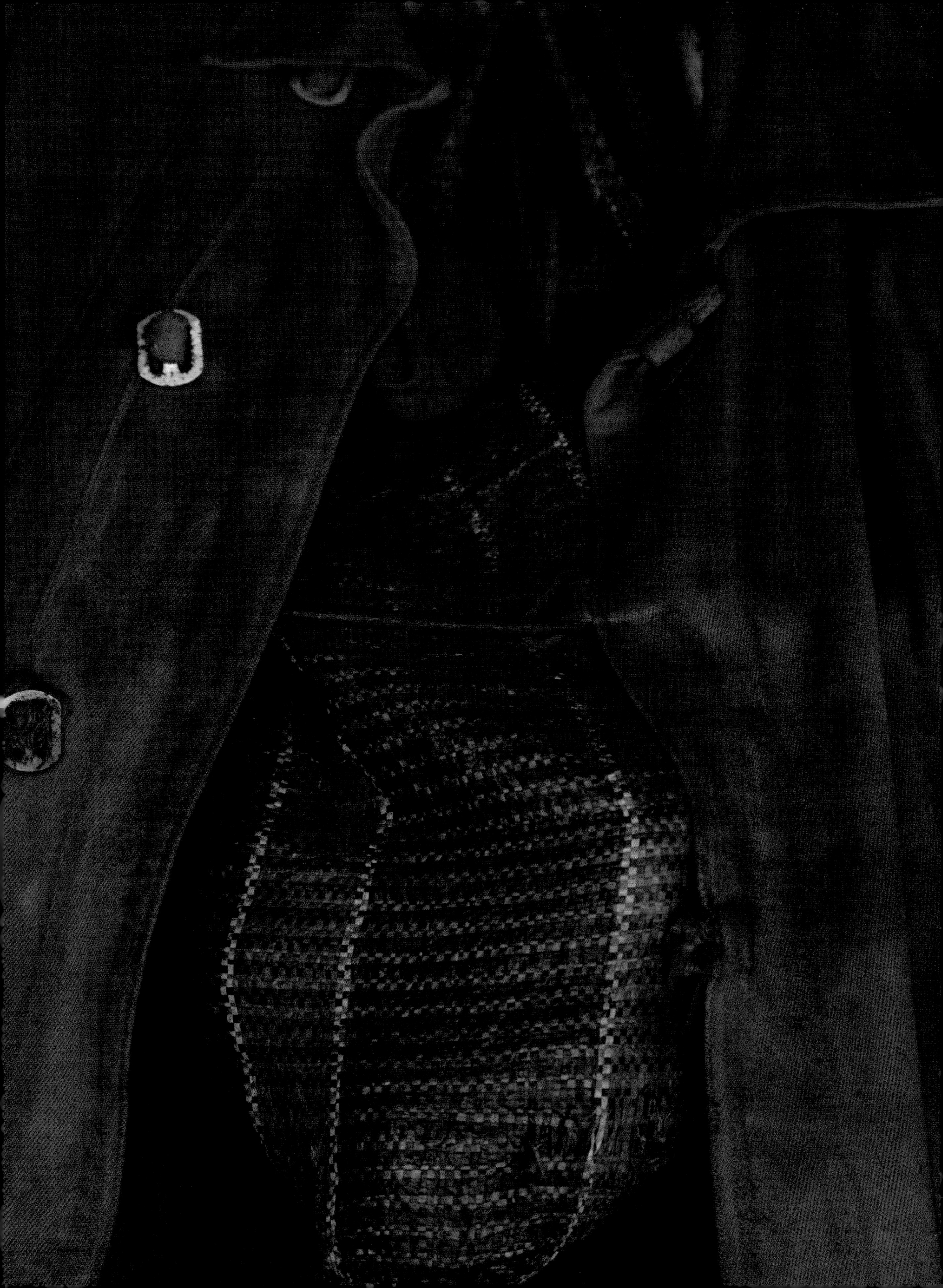

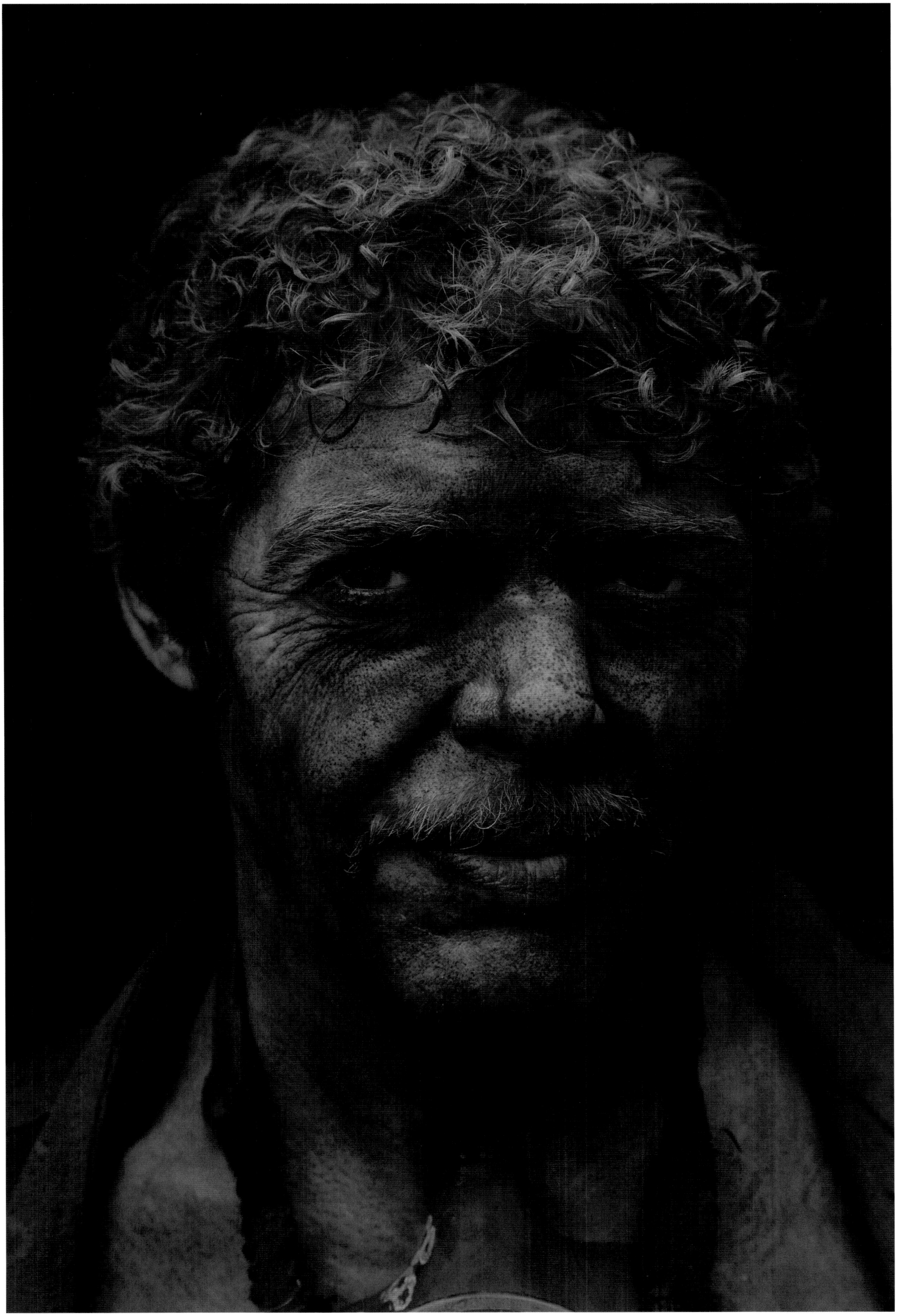

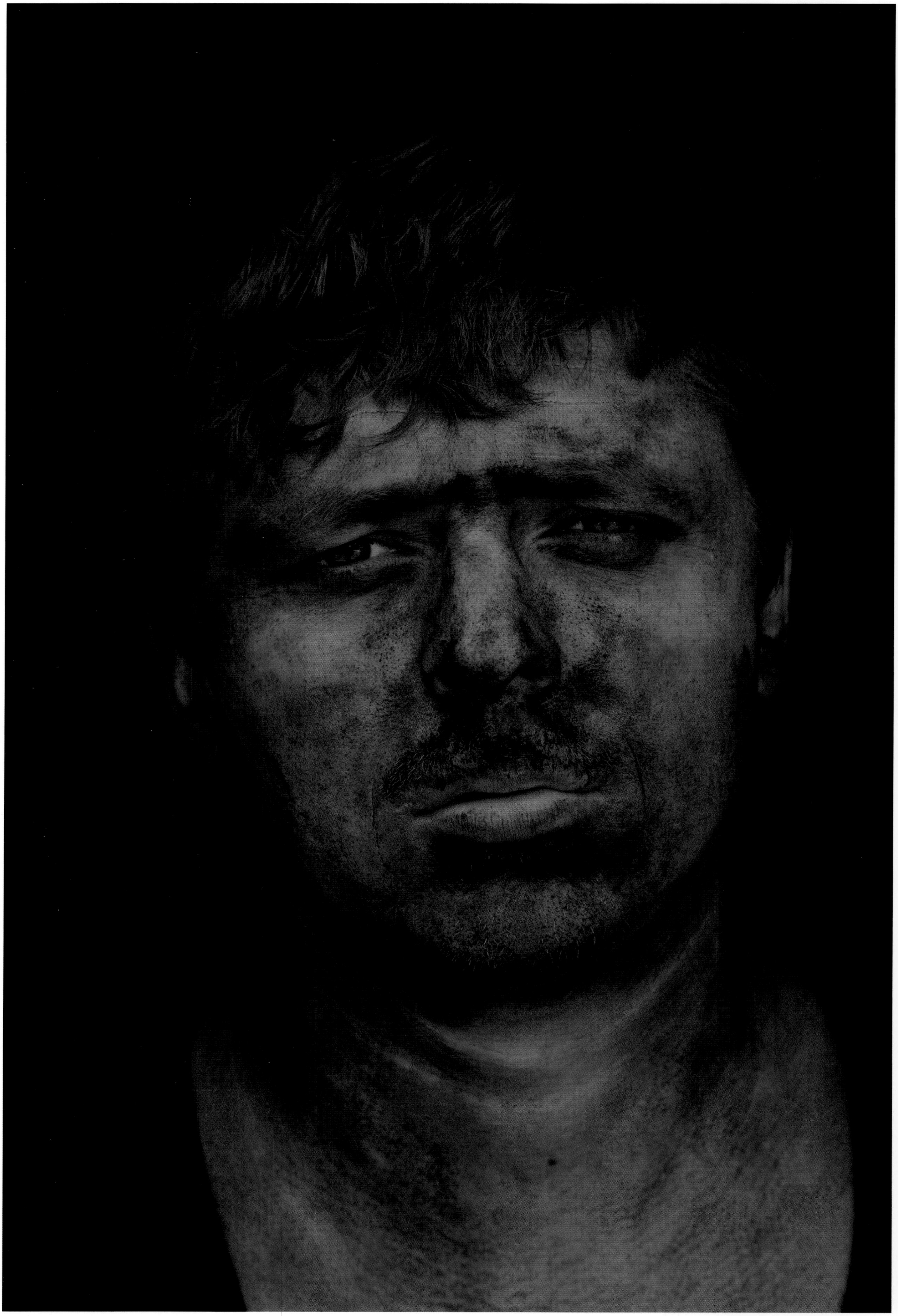

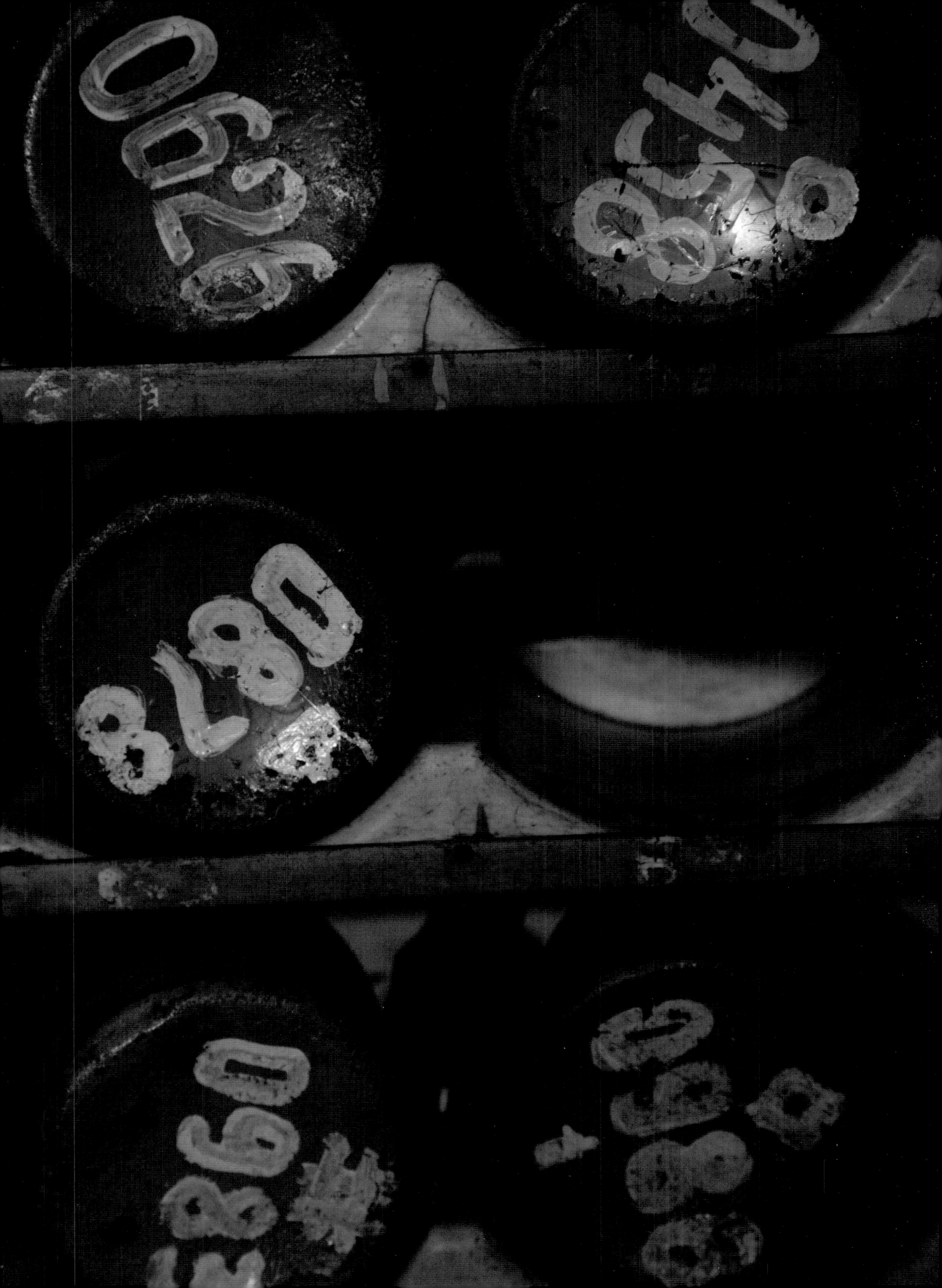

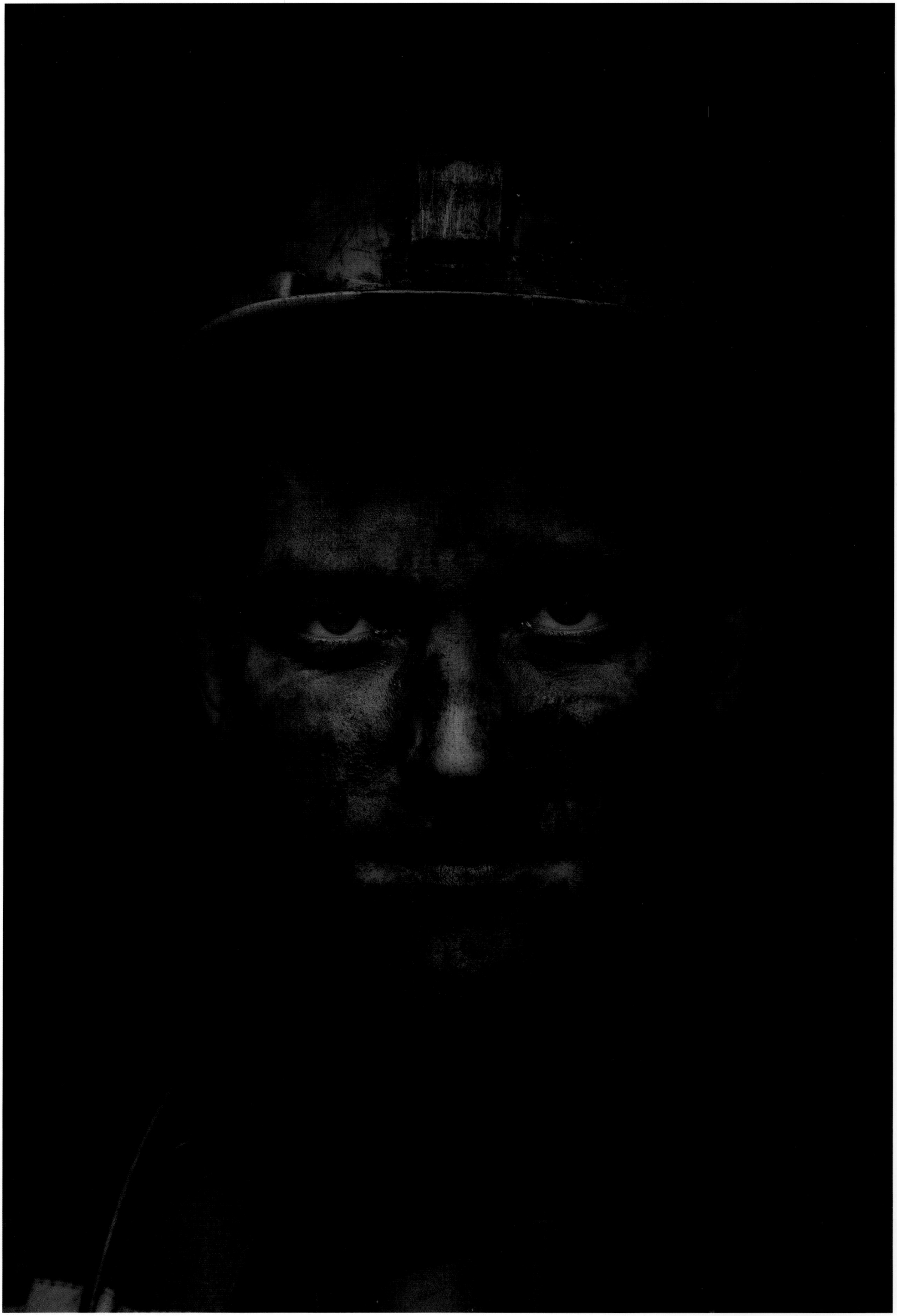

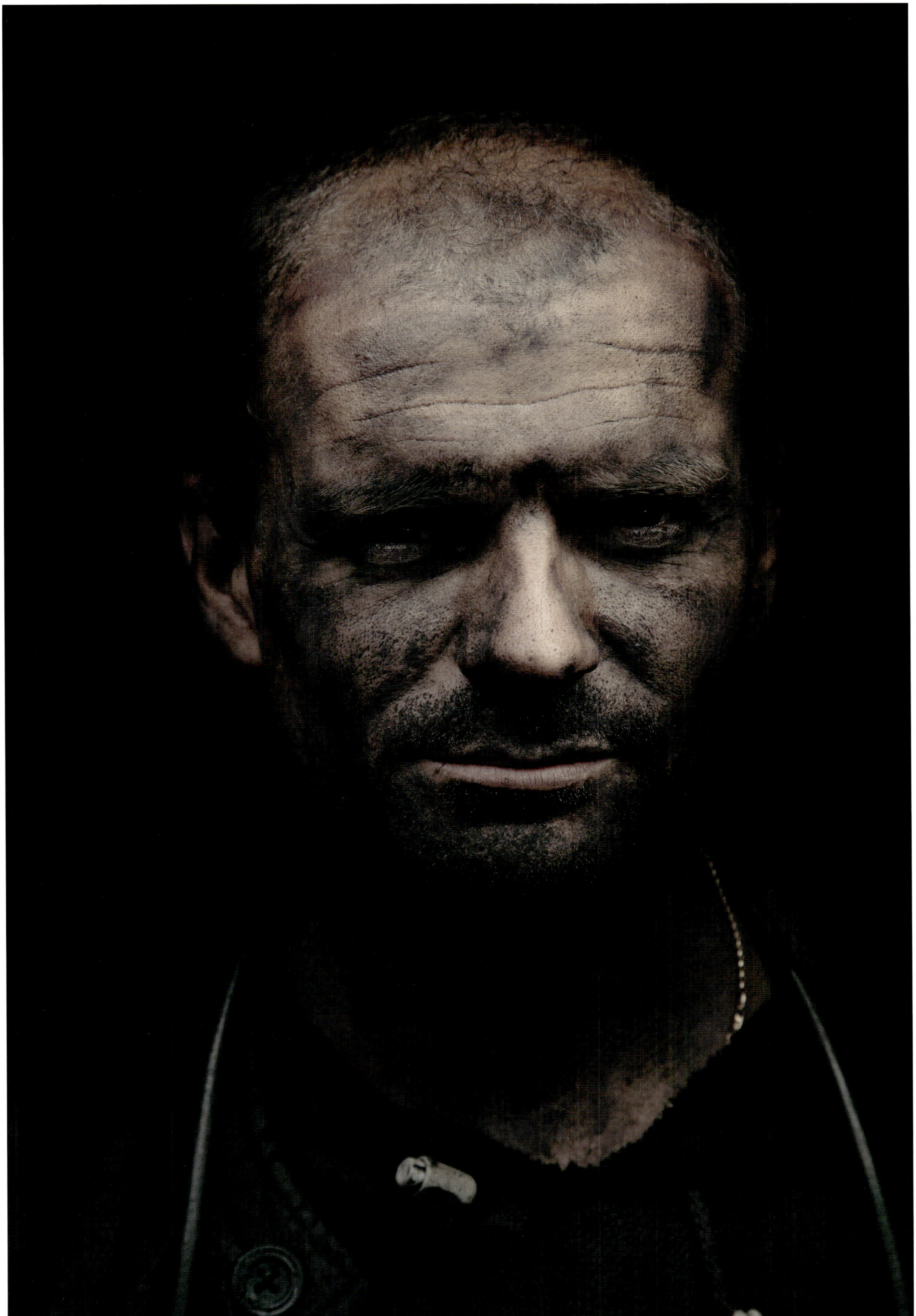

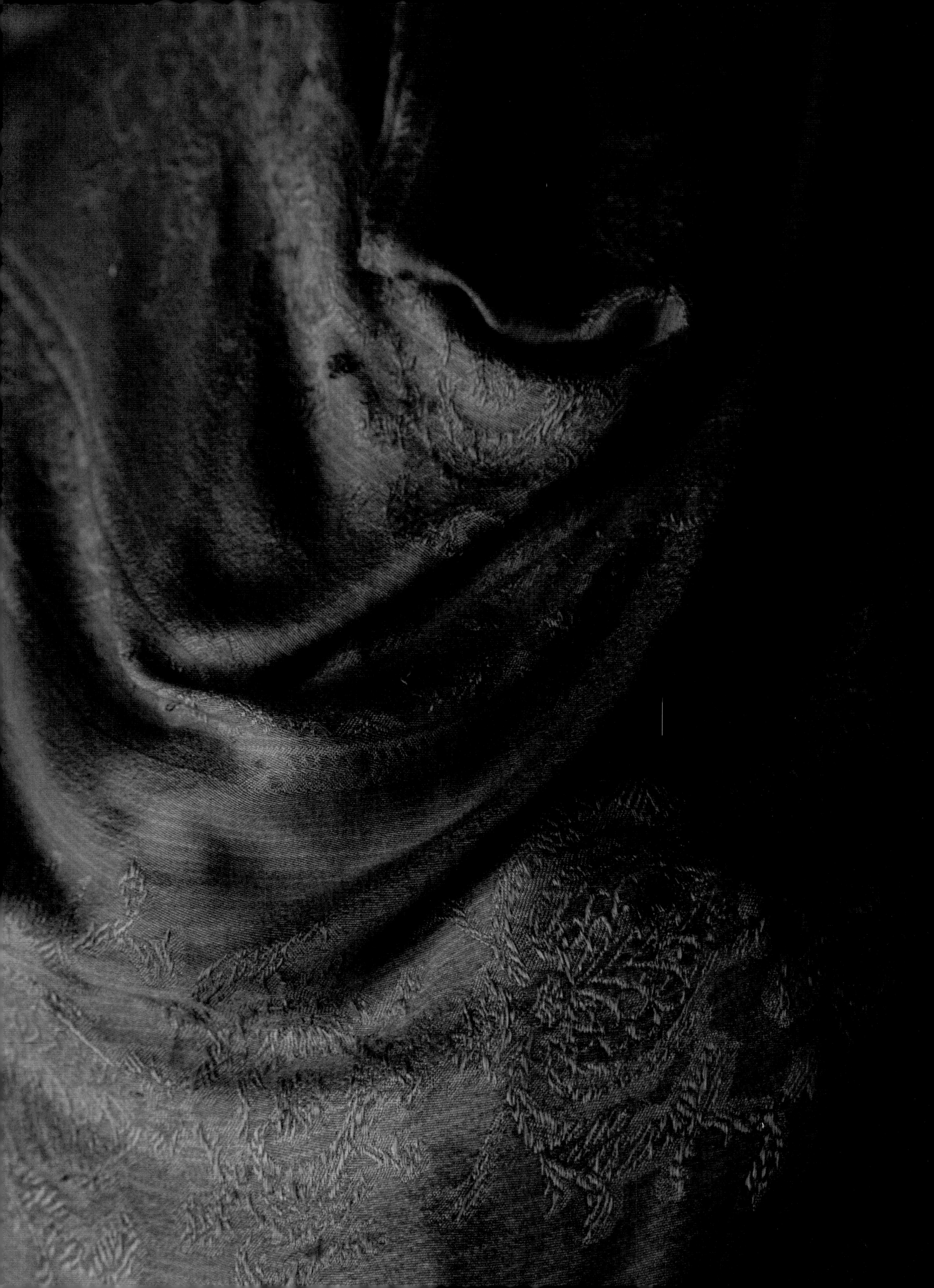

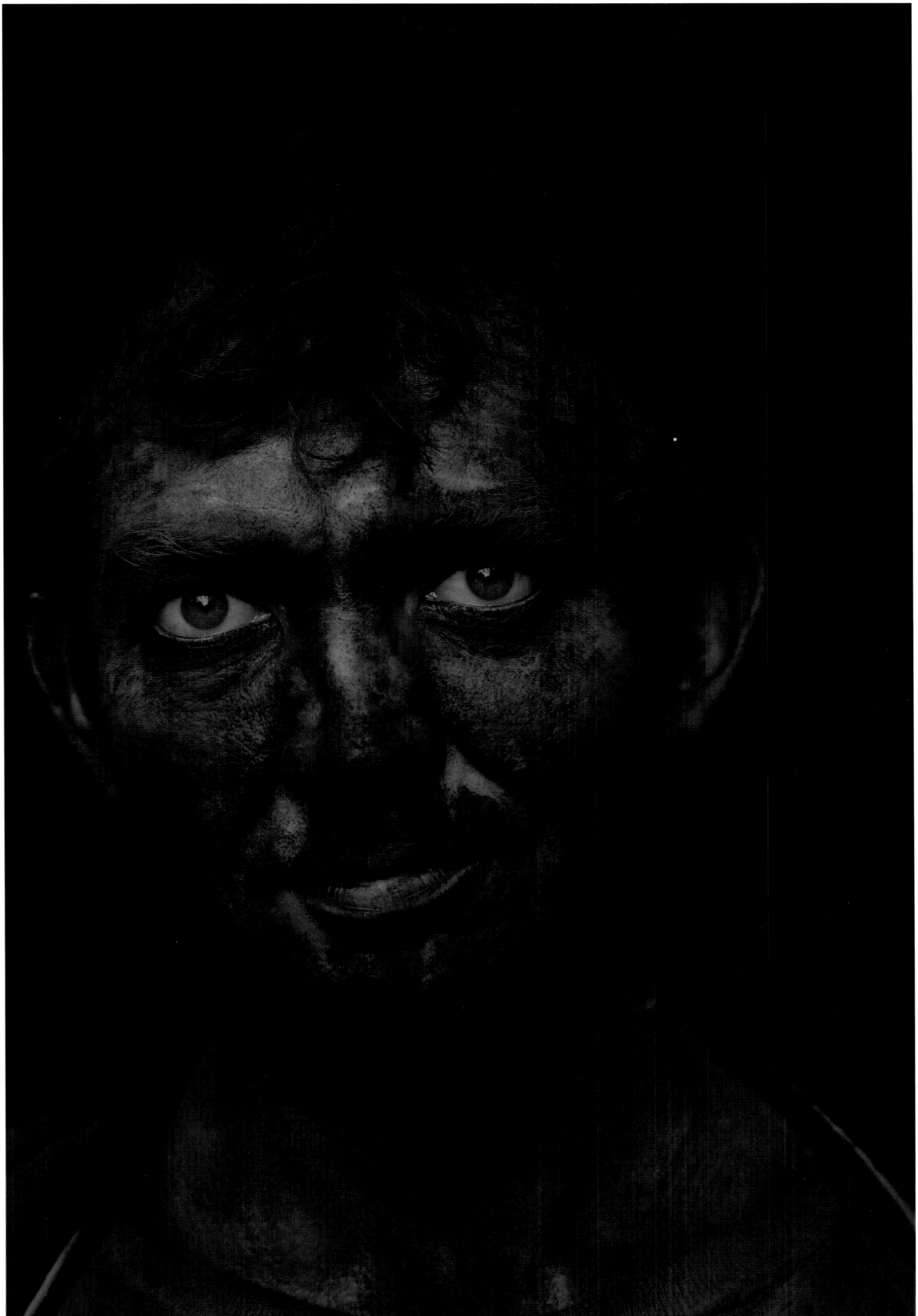

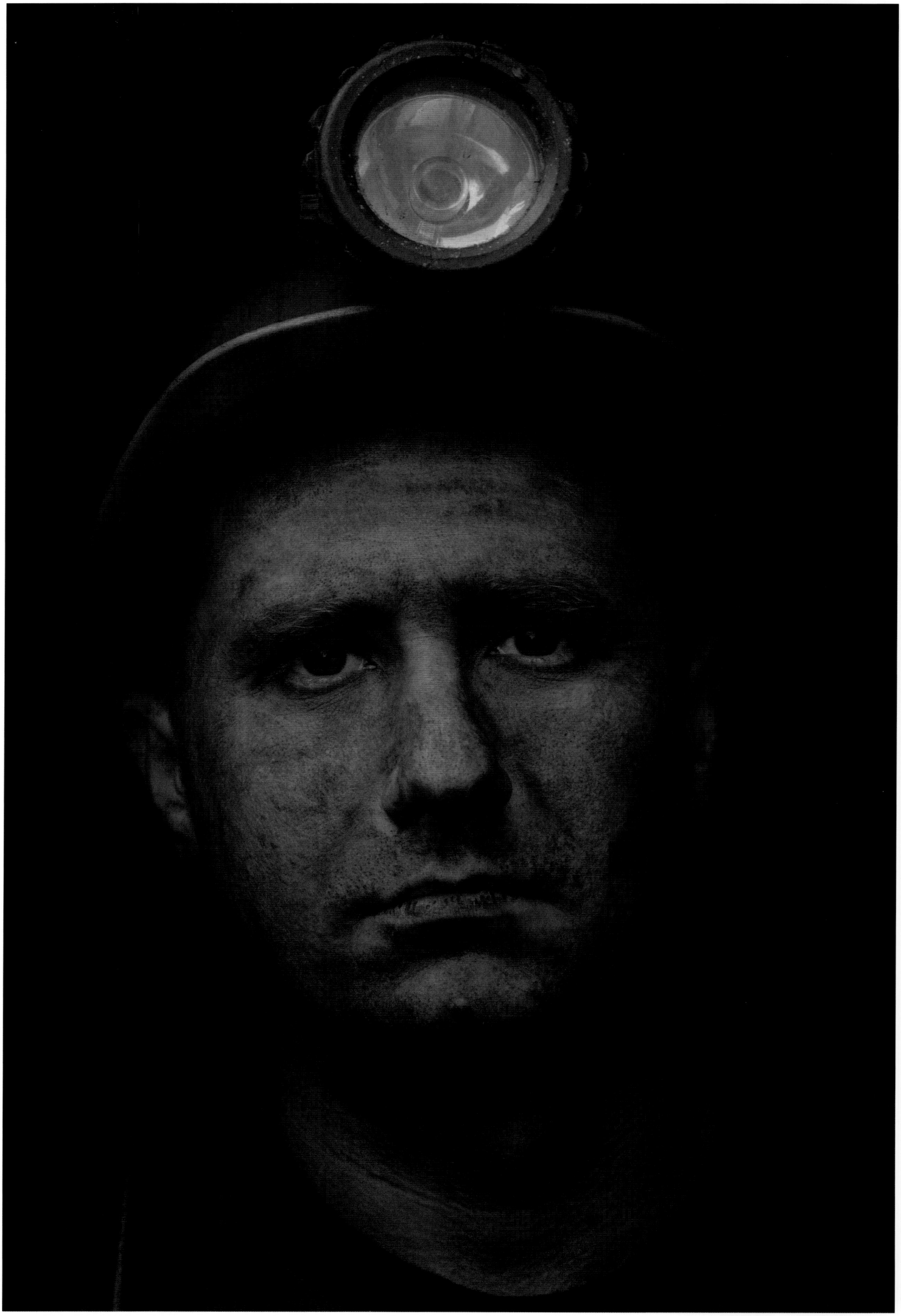

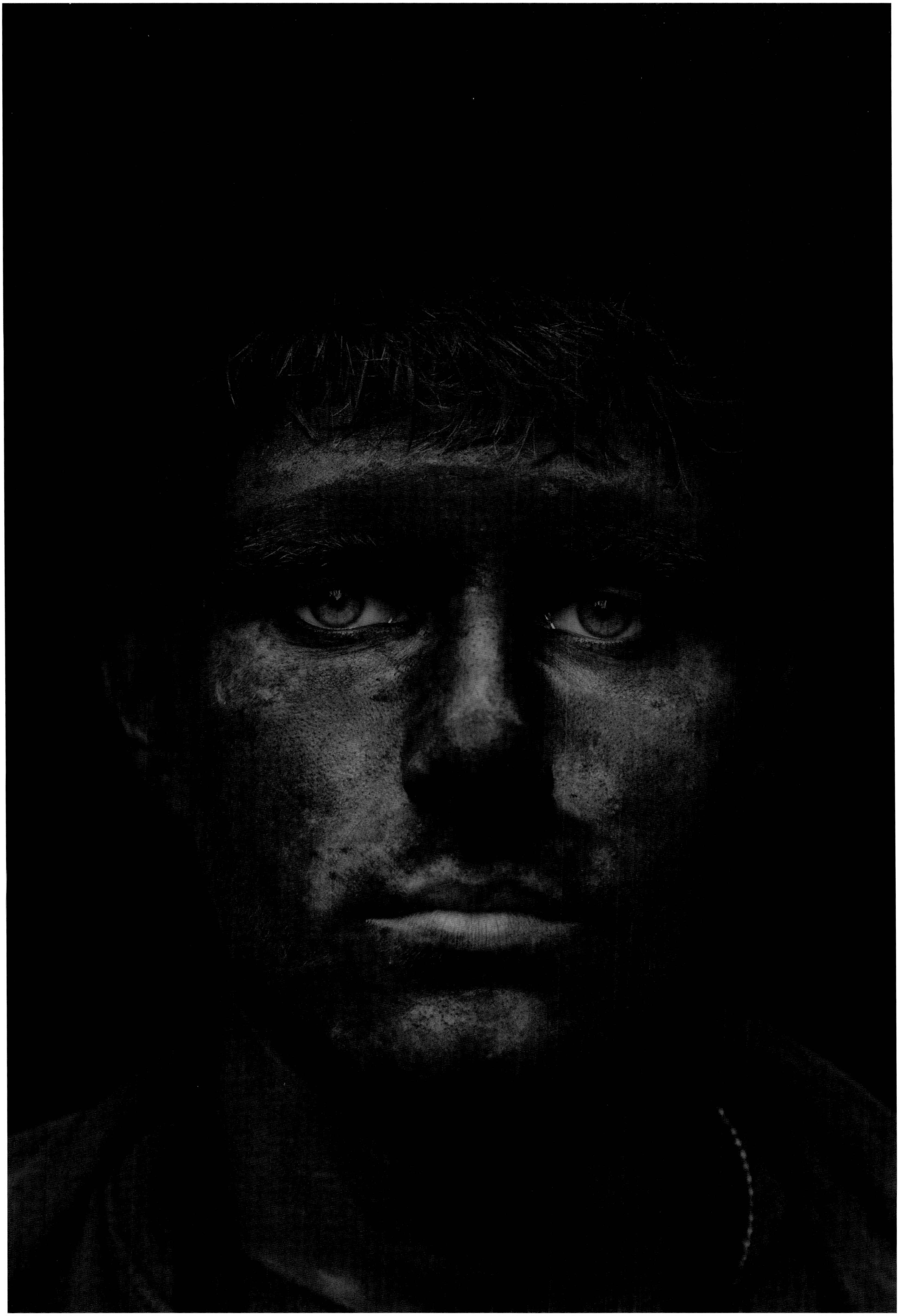

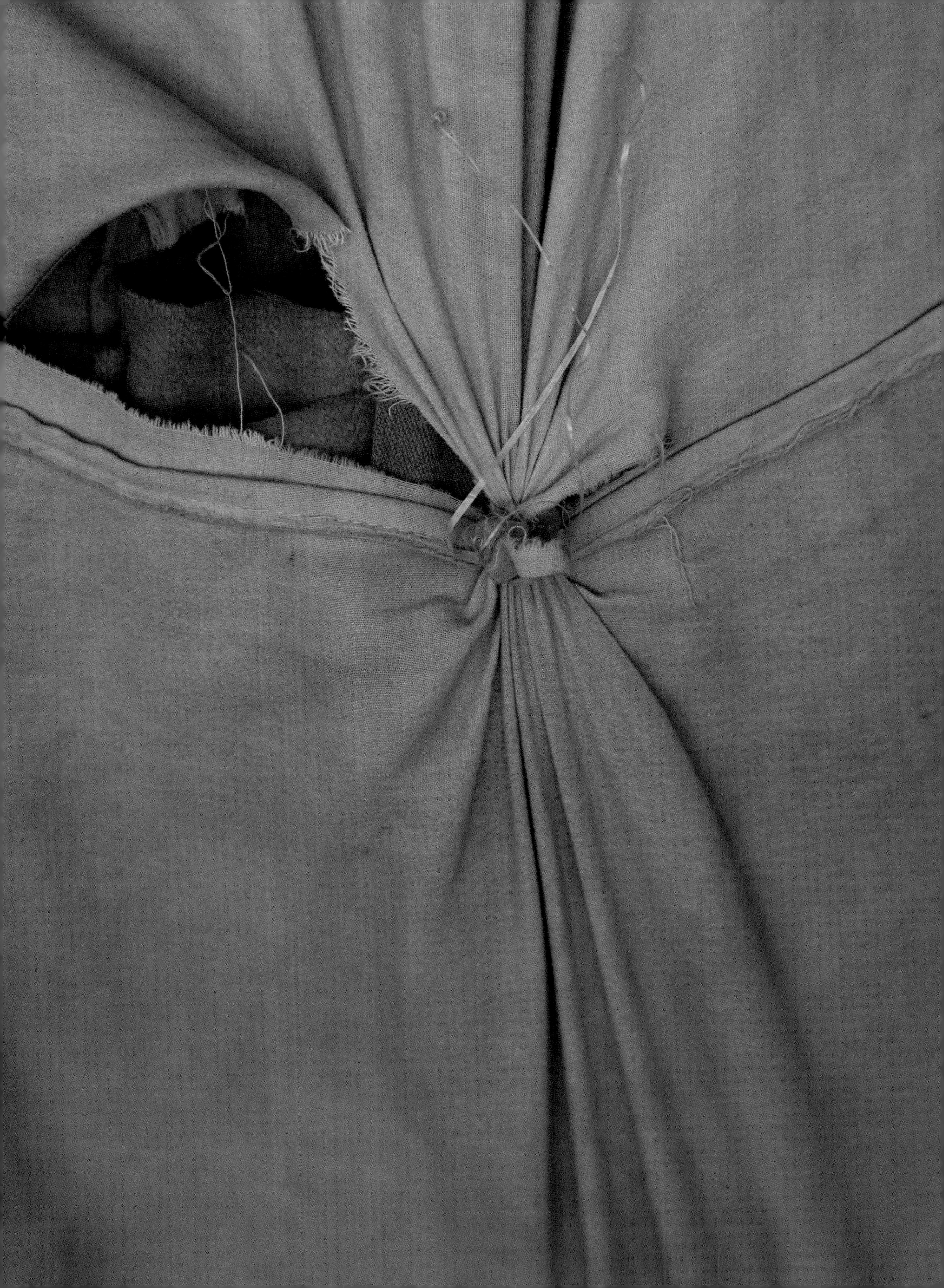

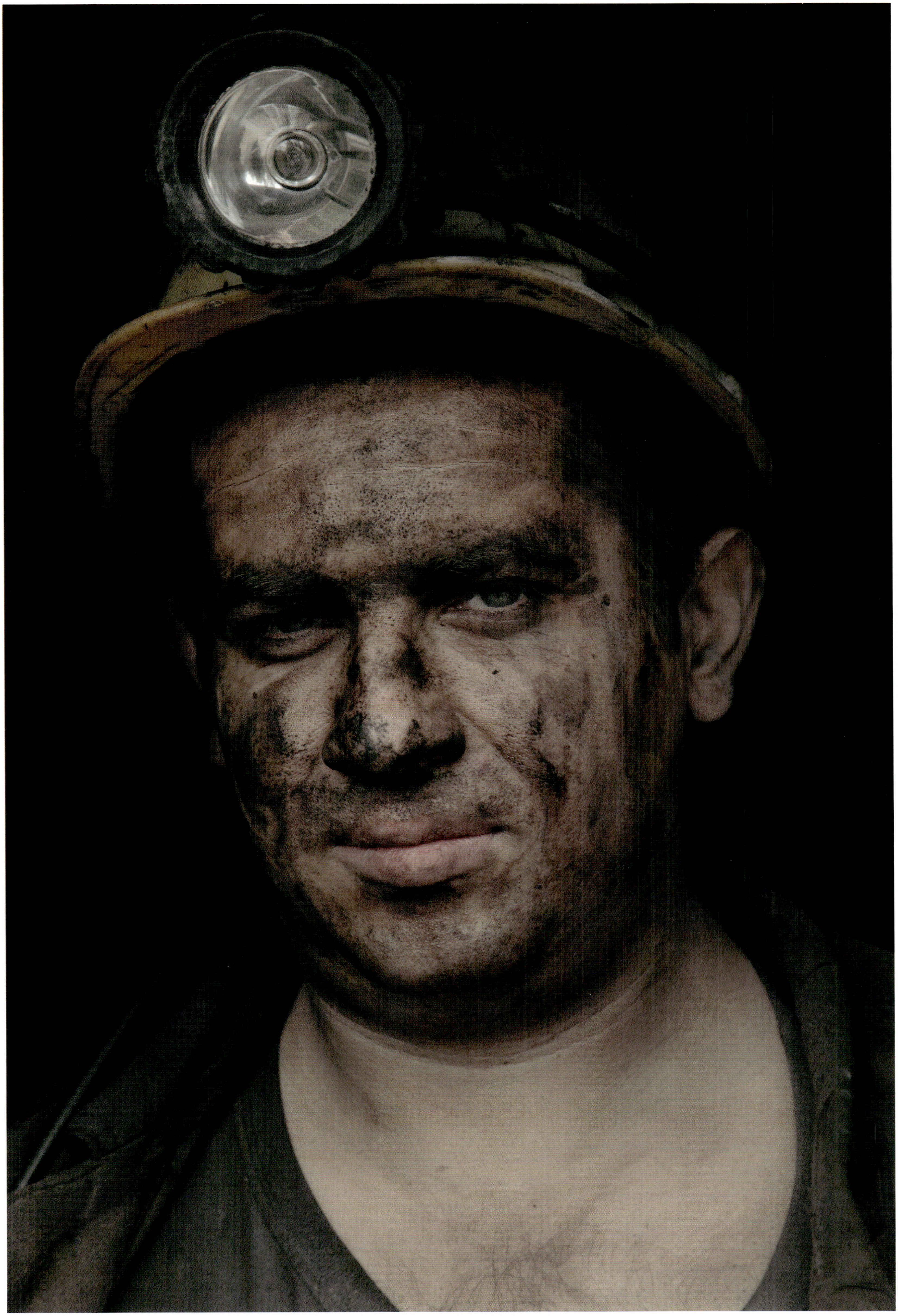

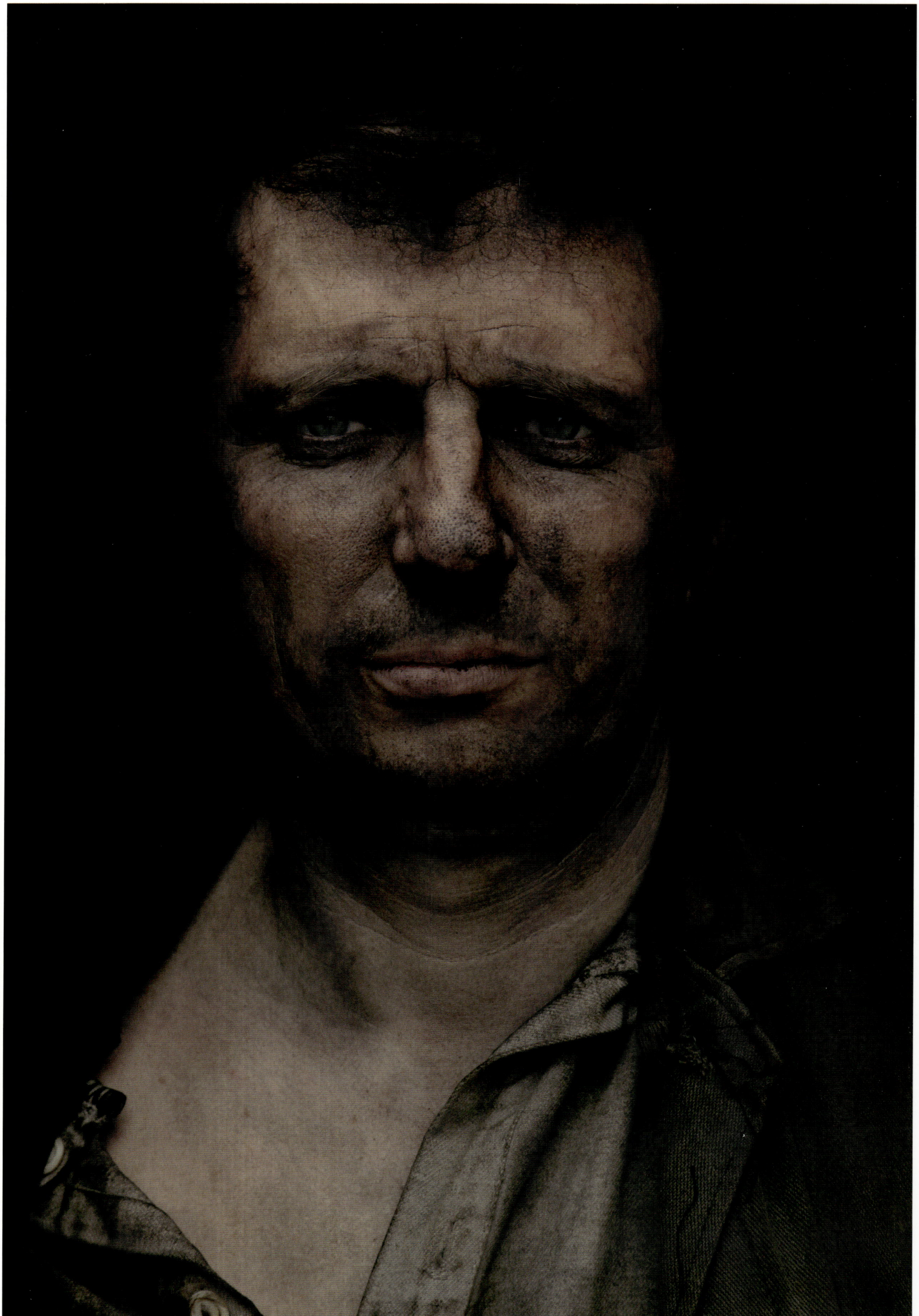

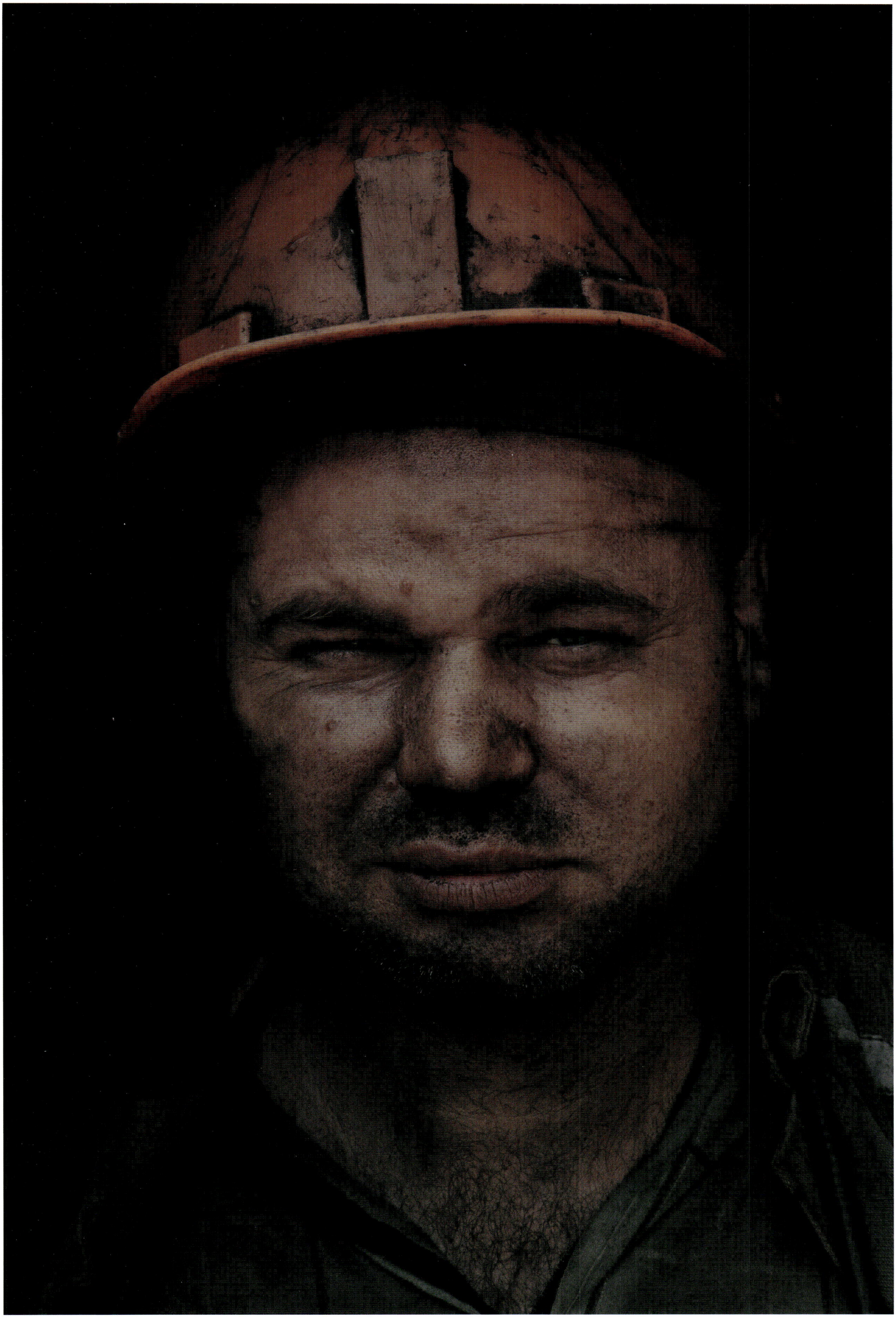

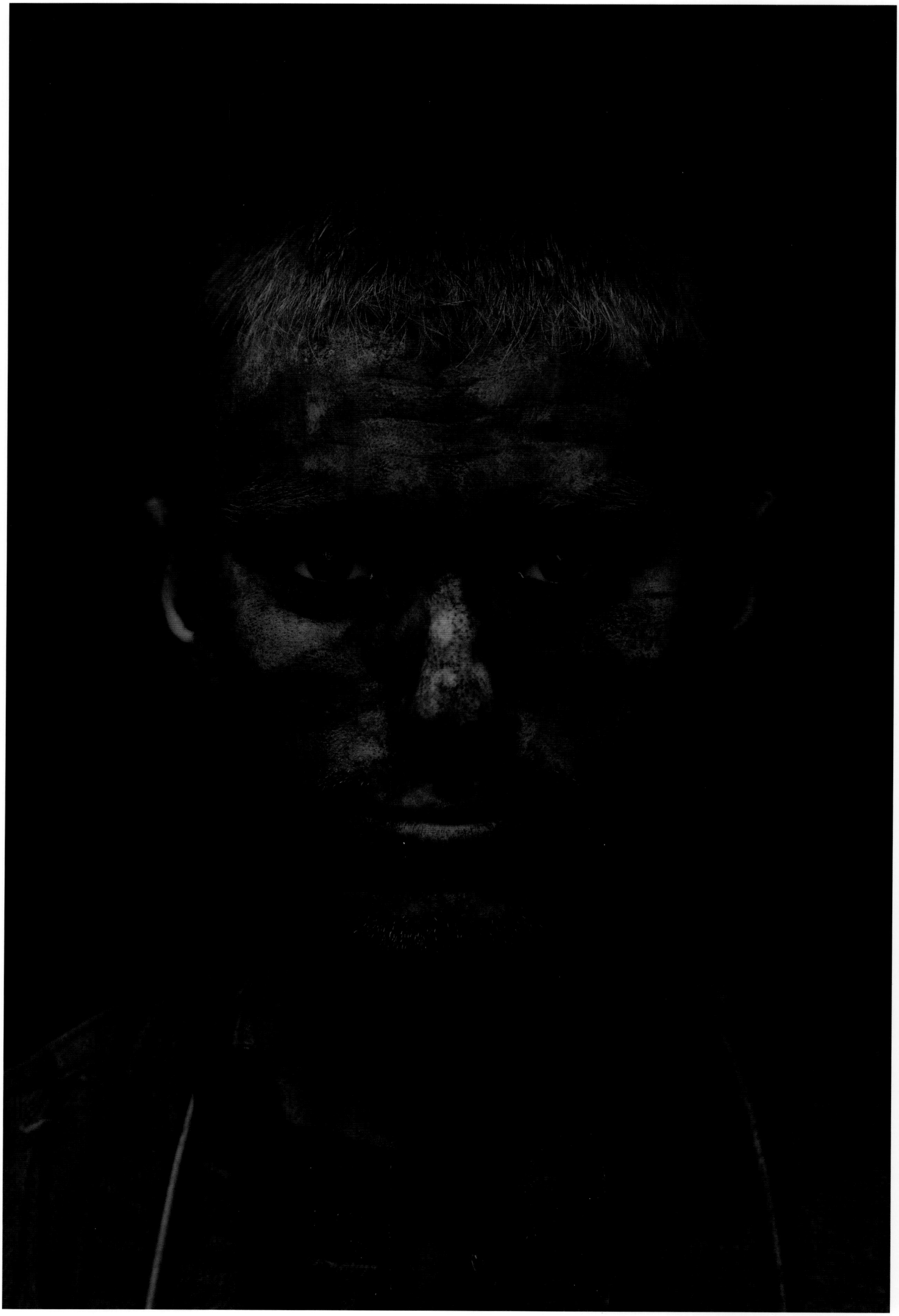

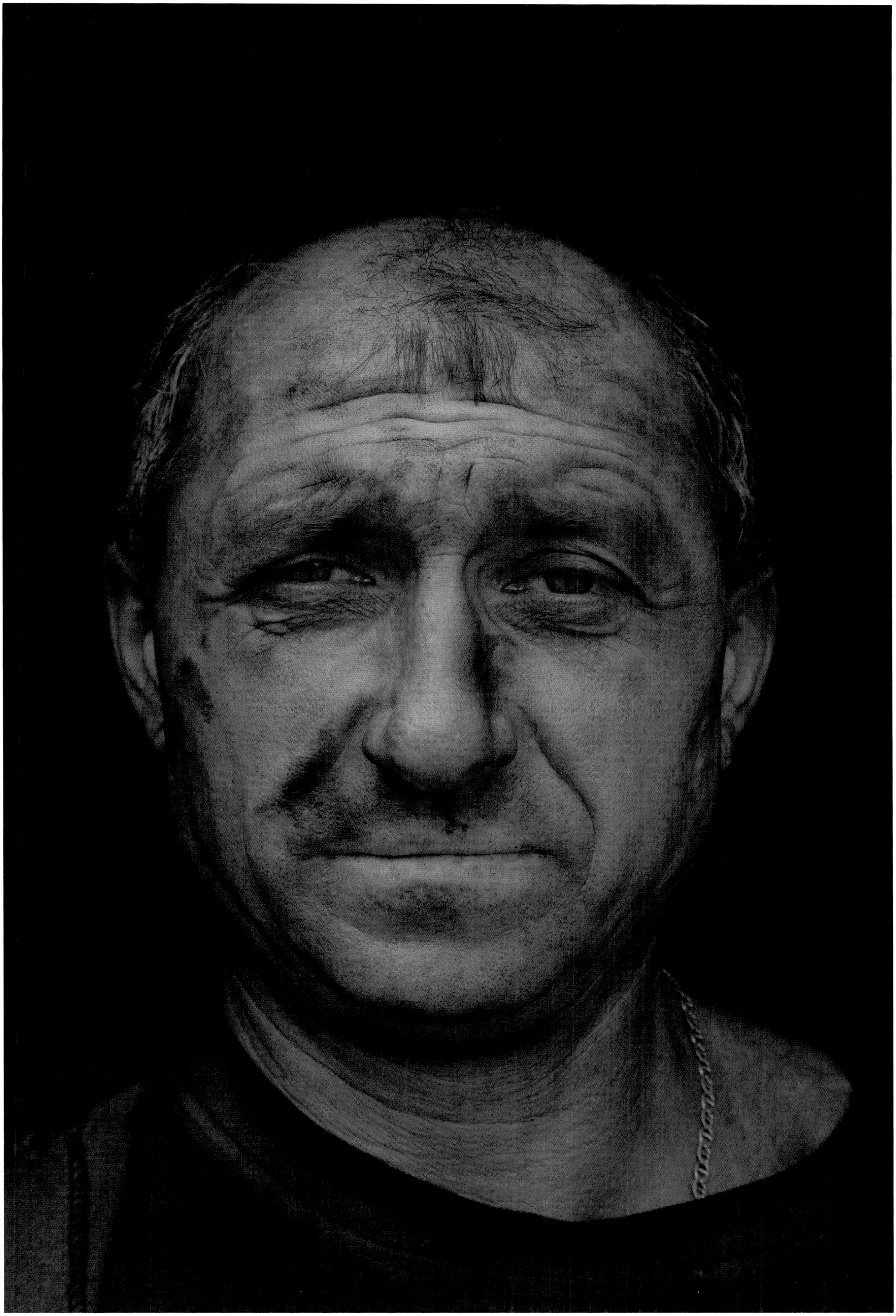

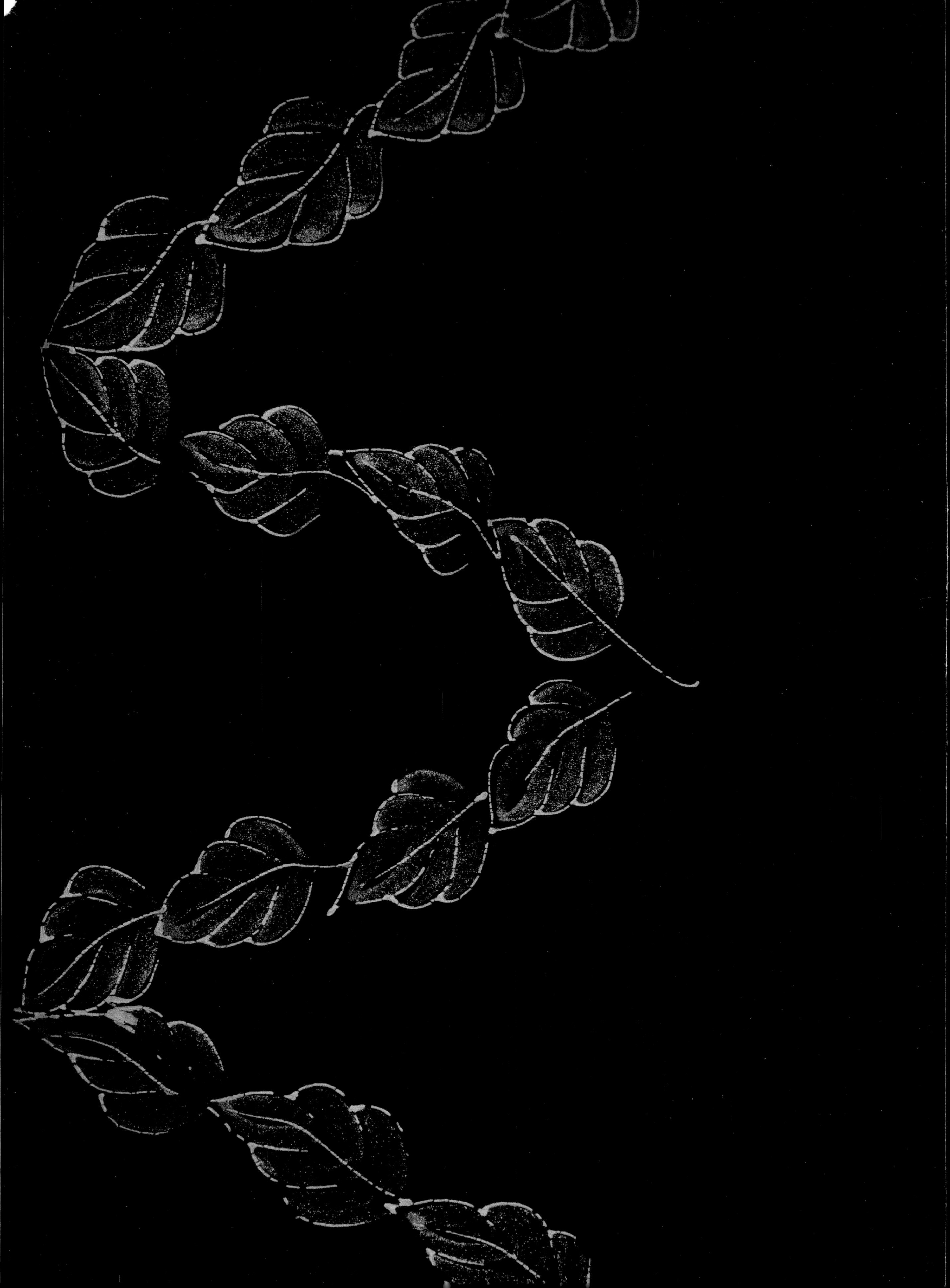

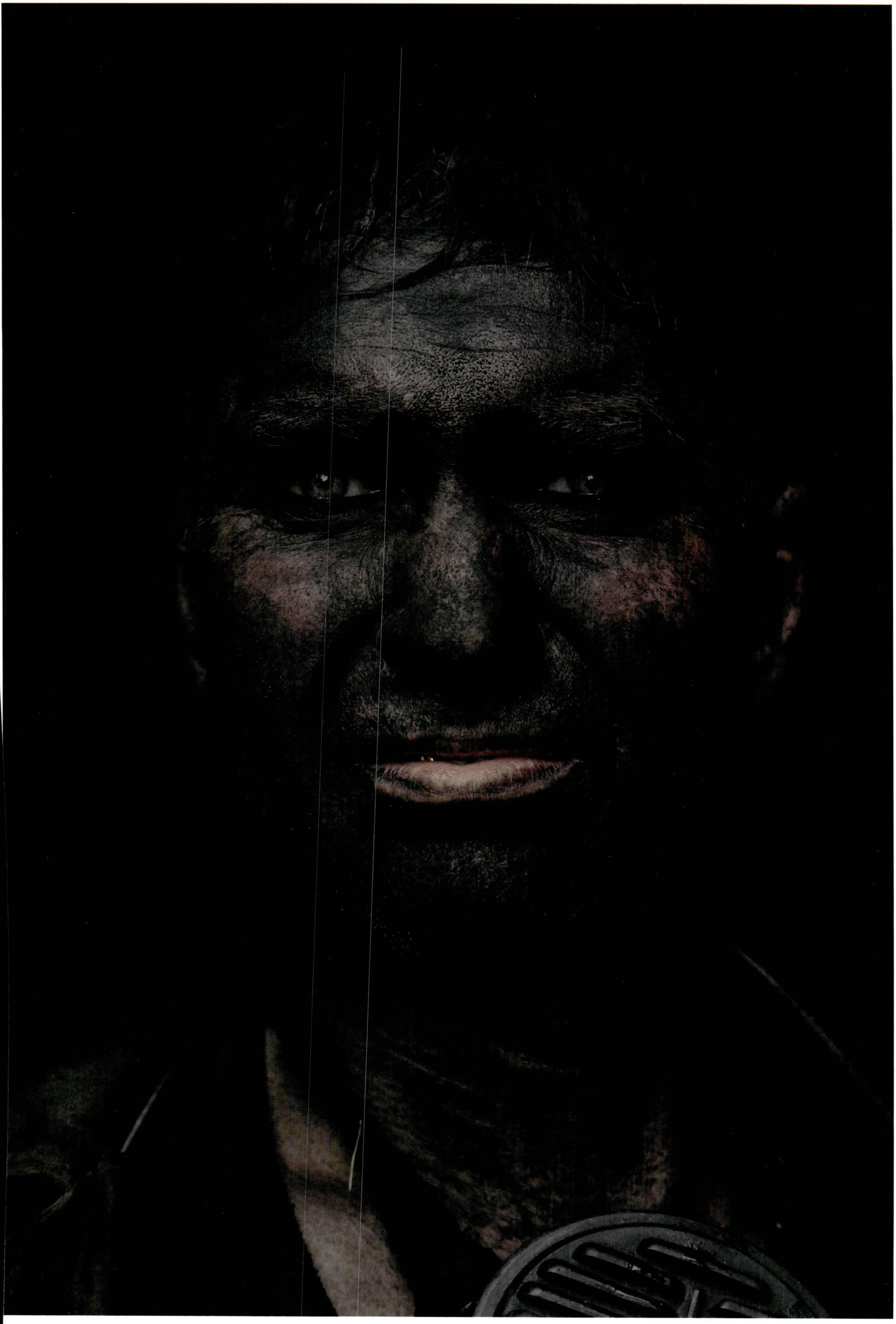

First edition published in 2017

Book design: Gleb Kosorukov
Separations by Steidl's digital darkroom

Production and printing: Steidl, Göttingen

Steidl
Düstere Str. 4 / 37073 Göttingen, Germany
Phone +49 551 49 60 60 / Fax +49 551 49 60 649
mail@steidl.de
steidl.de

ISBN 978-3-86930-689-6
Printed in Germany by Steidl